73rd Annual Meeting of the Labor and Employment Relations Association and LERA@ASSA Meeting 2021

Labor and Employment Relations Association Series

Online
5 – 8 June 2021
3 – 5 January 2021

ISBN: 978-1-7138-4521-8

Printed from e-media with permission by:

Curran Associates, Inc.
57 Morehouse Lane
Red Hook, NY 12571

Some format issues inherent in the e-media version may also appear in this print version.

Labor and Employment Relations Association (LERA)
University of Illinois at Urbana-Champaign
121 Labor and Employment Relations Bldg.,
504 E. Armory Ave.
Champaign IL 61820
USA

Phone: (217) 333-0072
Fax: (217) 265-5130

LERAoffice@illinois.edu

Additional copies of this publication are available from:

Curran Associates, Inc.
57 Morehouse Lane
Red Hook, NY 12571 USA
Phone: 845-758-0400
Fax: 845-758-2633
Email: curran@proceedings.com
Web: www.proceedings.com

LABOR AND EMPLOYMENT RELATIONS ASSOCIATION SERIES

Proceedings of the LERA 2021 Meetings

LERA@ASSA Meeting
January 3-5, 2021
Virtual

LERA 73rd Annual Meeting
June 5-8, 2021
Virtual

OFFICERS OF LERA FOR 2020-2021

PRESIDENT:
Adrienne Eaton, Rutgers University

PRESIDENT ELECT:
Wilma Liebman, Former Chair NLRB

PAST PRESIDENT:
Dennis Dabney, Kaiser Permanente

SECRETARY-TREASURER:
Andrew Weaver, University of Illinois at Urbana-Champaign

EDITOR-IN-CHIEF:
J. Ryan Lamare, University of Illinois at Urbana-Champaign

CHAPTER ADVISORY CHAIR:
William Canak, Middle Tennessee State University

EXECUTIVE DIRECTOR:
Emily Smith, University of Illinois at Urbana-Champaign

REGIONAL VICE PRESIDENTS:
Jim Pruitt, Kaiser Permanente (West Region)
Beverly E. Harrison, Arbitrator/Mediator (East Region)
Robert Chiaravalli, Strategic Labor and HR LLC (Mid Region)

EXECUTIVE BOARD MEMBERS:
Daniel Altchek, Saul Ewing Arnstein & Lehr LLP
Peter Berg, Michigan State University
Julie Farb, AFL-CIO
Janet Gillman, Oregon Employment Relations Board
Shannon Gleeson, Cornell University
Kati Griffith, Cornell University
Quenton Herbert, City of Baltimore
Erin Johansson, Jobs With Justice
Tamara Lee, Rutgers University
Kevin Legel, Ford Motor Co.
Deborah Mueller, CSEA Local 1000 AFSCME
Javier Ramirez, Federal Mediation and Conciliation Service
Christine Riordan,, University of Illinois at Urbana-Champaign
Jake Rosenfeld, Washington University in St. Louis

LEGAL COUNSEL:
Steven B. Rynecki, von Briesen, Purtell & Roper, S.C.

PAST PRESIDENTS OF LERA:

1948—Edwin E. Witte
1949—Sumner H. Slichter
1950—George W. Taylor
1951—William Leiserson
1952—J. Douglas Brown
1953—Ewan Clague
1954—Clark Kerr
1955—Lloyd G. Reynolds
1956—Richard A. Lester
1957—Dale Yoder
1958—E. Wight Bakke
1959—William Haber
1960—John T. Dunlop
1961—Philip Taft
1962—Charles A. Myers
1963—William F. Whyte
1964—Solomon Barkin
1965—Edwin Young
1966—Arthur M. Ross
1967—Neil Chamberlain
1968—George P. Shultz
1969—Frederick Harbison
1970—Douglass V. Brown
1971—George Hildebrand
1972—Benjamin Aaron
1973—Douglas H. Soutar
1974—Nathaniel Goldfinger
1975—Gerald G. Somers
1976—Irving Bernstein
1977—F. Ray Marshall
1978—Charles C. Killingsworth
1979—Jerome M. Rosow
1980—Jack Barbash
1981—Rudolph A. Oswald
1982—Milton Derber
1983—Jack Stieber
1984—Wayne L. Horvitz
1985—Everett Kassalow
1986—Lloyd Ulman
1987—Michael H. Moskow
1988—Phyllis A. Wallace
1989—Joyce D. Miller
1990—Robert B. McKersie
1991—James L. Stern
1992—Ernest J. Savoie
1993—George Strauss
1994—Lynn R. Williams
1995—Walter Gershenfeld
1996—Hoyt N. Wheeler
1997—Francine D. Blau
1998—F. Donal O'Brien
1999—Thomas A. Kochan
2000—Sheldon Friedman
2001—Magdalena Jacobsen
2002—John F. Burton, Jr.
2003—Paula B. Voos
2004—Marlene Heyser
2005—Stephen R. Sleigh
2006—David B. Lipsky
2007—Eileen B. Hoffman
2008—Anthony Oliver, Jr.
2009—Joel Cutcher-Gershenfeld
2010—Eileen Appelbaum
2011—Gordon Pavy
2012/13—David Lewin
2013/14—Lisa Lynch
2014/15—Martin Mulloy
2015/16—Bonnie Castrey
2016/17—Janice Bellace
2017/18—Harry Katz
2018/19—Kris Rondeau
2019/20—Dennis Dabney

PROCEEDINGS OF THE LERA 2021 MEETINGS

LABOR AND EMPLOYMENT RELATIONS ASSOCIATION SERIES
(Formerly the Industrial Relations Research Association)
Proceedings of the Annual Meeting
Annual Research Volume
Perspectives on Work Magazine

Inquiries and other communications regarding membership, meetings, publications, and general affairs of the Association, as well as notice of address changes, should be addressed to the LERA office.

LABOR AND EMPLOYMENT RELATIONS ASSOCIATION
University of Illinois at Urbana-Champaign School of Labor and Employment Relations
121 Labor and Employment Relations Building, MC-504
504 East Armory Avenue, Champaign, IL 61820
Telephone: 217/333-0072 Fax: 217/265-5130 E-mail: LERAoffice@illinois.edu
Website: http://lera.memberclicks.net

LABOR AND EMPLOYMENT RELATIONS ASSOCIATION SERIES

Proceedings of the LERA 2021 Meetings

LERA@ASSA Meeting
January 3–5, 2021, Virtual
(in conjunction with ASSA/AEA)

and

LERA 73rd Annual Meeting,
June 5–8, 2021, Virtual

J. Ryan Lamare, Editor-in-Chief

CONTENTS

LERA 73RD ANNUAL MEETING, JUNE 5-8, 2021, VIRTUAL

I.
PRESIDENTIAL ADDRESS

Facing Multiple Crises: A Quote, Two Other Presidential Addresses, and Two Books

Adrienne Eaton 1

II.
LERA BEST POSTERS SESSION

Hye Jin Rho, Presiding

Speak Up: Employees Voice in Regular Meetings Make Jack a Wise Boy

Eung Il Kim 7

The Impact of Collective Bargaining and Geographic Location on the Pay Rates of Professional Nurses: A Longitudinal Analysis (1970–1993)

Stephen Havlovic and Charles G. Smith 7

III.
LERA BEST PAPERS X: NATURE OF WORK PART A

Yea Hee Ko, Presiding

Retooling the Double-Edged Sword: Exploring the Impact of Functional Heterogeneity on Informal Cross-Functional Collaboration

Jeongrock Kim, Tasneem Omar Ava, Jalana Ellis, Daniela Febres, and Amanda Klavert 9

IV.
LERA BEST PAPERS XVII: WAGES

Michael Loconto, Presiding

Unions' Effects on Wage Inequality in Japan

Kyoko Suzuki 29

V.

2020 LERA LIFETIME ACHIEVEMENT AWARD

Lifetime Achievement Awards
Dennis Dabney, Awardee 53

LERA@ASSA MEETING, JANUARY 3-5, 2020, VIRTUAL

VI.

THE OLD GIG ECONOMY: THE EXTENT OF PAYROLL FRAUD IN CONSTRUCTION, ITS COST TO SOCIETY, AND APPROACHES TO ITS REGULATION

William Canak, Presiding

Discussant Comments
Matthew Capece 55

VII.

LERA ANNUAL REPORTS

Executive Board Meeting Virtually, February 5, 2021 61
Executive Board Meeting Virtually, June 1, 2021 67
General Membership Meeting Virtually, June 6, 2021 71
Audit Report for 2020 75
INDEPENDENT AUDITOR'S REPORT 78-79
FINANCIAL STATEMENTS
Statements of Financial Position 80
Statement of Activities – For the Year Ended December 31, 2020 81
Statement of Activities – For the Year Ended December 31, 2019 82
Statement of Functional Expenses – For the Year Ended December 31, 2020 83
Statement of Functional Expenses – For the Year Ended December 31, 2019 84
Statements of Cash Flows 85
NOTES TO FINANCIAL STATEMENTS 86-97
Sustaining Sponsors and Organizational Members 99
Alphabetical List of Authors 100

I. Presidential Address

Facing Multiple Crises: A Quote, Two Other Presidential Addresses, and Two Books

ADRIENNE EATON
Rutgers University School of Management and Labor Relations

Editor's note: LERA president Adrienne Eaton delivered her presidential address at the virtual LERA 73rd Annual Meeting on June 7, 2021.

I've been thinking about this talk for a long time and struggling with what to say in this moment and given the past year. I had originally been thinking I would focus this talk on democracy, and I spent some time reading the works of various political scientists and philosophers on the linkages between political and economic democracy—something I've long been interested in. But as I shifted my reading into some new areas and thought more about all the events of the past year, I headed in a different direction.

Starting with the idea that we, as members of the LERA community, as professionals in the field of employment relations, and as human beings, are facing numerous crises, I was inspired in part to think about multiple crises by the introduction to the 2021 LERA research volume, *Revaluing Work(ers): Toward a Democratic and Sustainable Future*, written by my Rutgers colleagues Toby Schulze-Cleven and Todd Vachon.

Foremost in our minds—the thing that continues to keep us apart for this conference — is the COVID-19 pandemic. The pandemic has upended work, the institution of central concern that LERA revolves around. With the shutdowns in March 2020, the pandemic crisis quickly became an employment and work crisis. As I said to a student newspaper reporter around that time, people were working from home (something many had never done before), working unsafely, or not working.

> The impact of both the pandemic and climate change have been and will continue to be felt unequally and have exacerbated inequalities of multiple kinds, both within the United States and globally.

And many of those who were not working and maybe are still not working were wondering (or still are) about where their next meal was going to come from. I think here especially of the undocumented among us who have little or no access to the social safety net. Here are some quotes from undocumented workers in New York City, most from spring 2020.

> Well, in the case of my family, we are very scared because we had the loss of my brother who died from the cv. And everyone is scared by this pandemic that we have had to live in isolation, without work, without money and with great fear.[1]
>
> It is impacting us financially, emotionally—there is sadness, worry, fear, and anguish among my children because of the fact that we, their parents, could die. We as parents are concerned about financial insecurity and not knowing when we will return to work.

> I'm worried because my dad lost his job, and my mom is the only one working. She's working at a laundromat. She's risking her health to work, and I'm scared my parents will get sick because they also are undocumented.
>
> There were days I couldn't sleep. To be honest with you, we had nothing.[2]

Returning now to my own voice, under the former administration, OSHA refused to do anything about workplace safety, though some states did, including my own. The federal government expanded the definition of eligibility for unemployment insurance and increased the payments; many states, either purposefully or as a result of poor IT systems or both, had enormous trouble implementing it. Healthcare workers improvised protective equipment and new clinical spaces; many worked unsafely, and some died—3,600 in the United States alone, according to estimates by Kaiser Health News and *The Guardian*.

All these topics and more have provided challenges and lessons in the world of practice—and a rich world to study, however ghoulish that might sound. And research is already emerging from those with the time and energy to do it. Perhaps this entire talk should be dedicated to work and the pandemic, but I thought we might all be a bit tired of it, honestly.

Then there is climate change, an existential threat to us all and one that, as demonstrated in last year's conference theme around sustainability, touches many areas in which we work—from green jobs to just transitions, to immigration and immigrant workers. It's a hard topic to center in the day-to-day work that many of us do, yet it looms over everything.

As an aside, I want to mention my favorite quote that links climate change and the word of labor and employment relations. I first heard it from Sharan Burrow, the general secretary of the International Trade Union Confederation. But in googling around, I learned that someone named Judy Bonds is often credited with saying it first. She was a coal miner's daughter and an environmental activist in West Virginia, which can help you imagine the context for that quote. She died of cancer in 2011.

The quote is "There are no jobs on a dead planet." And by the way, I've also seen the quote adapted to "There is no business on a dead planet." And I suppose we could say there are no publications and no tenure on a dead planet either. Our field needs to grow in its engagement with this crisis, no matter the particular work we each do.

Then there are the interrelated crises of inequality, racism, and democracy. The impact of both the pandemic and climate change have been and will continue to be felt unequally and have exacerbated inequalities of multiple kinds, both within the United States and globally, as we can now see with the vaccine rollout. I believe our field has been centrally concerned with some kinds of inequality since its founding, the inequality of power between workers and capital or managers, if you prefer—what we might call class inequality.

> Personally, I am very drawn to the argument that we actually cannot and should not separate race and class.

I think we've been less focused on racial inequality, and we've been called to task for that by my colleague Tami Lee and her co-author Maite Tapia, among others, of course. I strongly urge you to read their paper, "Confronting Race and Other Social Identity Erasures: The Case for Critical Industrial Relations Theory" in the May 2021 special issue on theory in *Industrial and Labor Relations Review*. It may raise more questions for you than give answers, but that's a good thing, especially for the field. I'm not sure I agree with everything that Tami and Maite say in their paper, and I don't think I completely understand it, but I know we need to absorb the argument, sit with it, and think about what it means for our own work—

whether that work is research and teaching or practice in the field. I am proud that the 2022 LERA research volume is being edited by Maite and Tami along with two other Rutgers faculty members, Naomi Williams and Sheri Davis.

Personally, I am very drawn to the argument that we actually cannot and should not separate race and class. I've been very interested in the work of Ian Haney López in this regard and have been trying to find time to read his latest book, *Merge Left*. Haney López is a law professor at UC-Berkeley, where he studies and writes on politics. He's the author of a book on racial "dog whistles" (*Dog Whistle Politics*). I'm going to assume you know what those are. The subtitle of his newer book is *Fusing Race and Class, Winning Elections, and Saving America*. I highly recommend the book, which—of interest to this audience—starts with the story of a visit he paid to an annual retreat of a national union's leadership in Florida.

He argues that racism is fundamentally a class weapon, one used by moneyed interests to misdirect white working-class people about who or what is to blame for the rampant inequality they experience along with working class people of color. In a way, this message isn't terribly new, but—and I confess here I'm not that deep into the book—in this case, it's rooted in a lot of polling and survey research, and Haney López is all about using his findings to produce better political and organizing messaging.

Turning to inequality and its relationship to threats to democracy, I was also inspired by reading Kathy Thelen's presidential address to the American Political Science Association, titled "The American Precariat: U.S. Capitalism in Comparative."[3] It is quite interesting, though not surprising for those of you who know Professor Thelen's work, that her talk is really centered on topics that we focus on at LERA—what she labels "atypical employment," which some of us would prefer to call "informal employment" or "precarious employment."

She presents a series of graphs. Some of her graphs are focused just within the United States, showing the nature of class inequality. One graph shows access to paid sick leave, and I noticed it in part because that issue became such a huge issue during the pandemic. The left-hand side shows the decline over time by different occupational groups, with the lowest access always for production workers and the steep decline among technical and clerical workers. Another chart shows access to paid sick leave by income quartile, which is pretty self-explanatory.

The other set of charts in the piece shows a series of comparisons between Europe and the United States, mostly relying on Organisation for Economic Co-operation and Development data. I'm sure these will not be shocking to many of you. I can say that these kinds of comparisons are often shocking to Rutgers students. One chart shows statutory paid time off, for both vacations and holidays. The United States is on the right with zero. The second chart is not from Thelen's talk. I found it in a February 2016 article, "How U.S. Employee Benefits Compare to Europe's," in the magazine *Fast Company*. Yeah, we're over there on the right again—meaning least-generous unemployment insurance, parental entitlements, annual leave/holidays, sick pay. Again, these benefits rose in importance during the pandemic and continue to be so. For instance, in a recent Kaiser Family Foundation poll on vaccine hesitancy, 21 percent of unvaccinated workers said they'd be more likely to get the vaccine if their employer provided paid time off to both get the vaccine and recover from any side effects.[4]

I also want to mention here a recent NBER study summarized in a May 28 article in the *New York Times*.[5] This study shows how women's labor force participation fell compared to men's in 18 out of 28 countries in North America and Europe, with the greatest gaps in the United States and Canada. These gaps were a function of school closures (which were longest in the United States and Canada) and lack of robust childcare systems, but also gaps in other labor market policies, such as the use of furloughs versus unemployment and the differential impact of telecommuting on women's and men's work at home.

> The issues facing women workers are pervasive ..., creating inequities that are frankly without good answers or at least ones most of our institutions can easily afford.

I am not going to speak to gender again in this talk, but I want to highlight here how important a gendered lens can be to our research, which contrasts what the NBER study authors call the pandemic-induced "she-cession" versus earlier "man-cessions" and the different policy implications of these events. I want to note also that the issues facing women workers discussed in the study are pervasive in higher education, creating inequities that are frankly without good answers or at least ones most of our institutions can easily afford.

The third chart I want to discuss is again from Thelen's paper. It presents union density and coverage, a perennial favorite for us. The data are a little old, but I don't think anything has changed fundamentally—the United States is close to the edge again, although this time it's the left edge, only close because our union density is higher than France's. Union coverage, of course, tells a different story.

Thelen sums up: "All the rich democracies are experiencing a shared problem of growing contingency and precarity. However, it seems fair to say that the problem of precarity presents itself with special intensity in the United States." She also links political democracy to economic democracy and the need for political science as a discipline to study macro political economy and not just voting behavior and individual attitudes.

Thelen's concern with democracy brings me to another LERA presidential address, Paula Voos's to LERA—well, actually, to IRRA—in 2003. Paula's address was on democracy and industrial relations. Her first sentence reads: "Political democracy is thriving in the world." Hmm. Things have changed. To be fair, she goes on to say that "democracy may be doing well in the world, but it is not doing well in the United States," citing low rates of voting and particularly lower rates for those with lower incomes.

Both Voos and Thelen argue that income and wealth inequality constitute threats to democracy, with Voos emphasizing the ways unions can strengthen democracy not just by encouraging voting but also through reducing inequality. Schulze-Cleven and Vachon also discuss the links between precarity and status anxiety and openness to right-wing populism in their LERA research volume introduction, as does Guy Standing in the very title of his well-known book, *The Precariat: The New Dangerous Class*. My point here is that the concerns of our field, focused on employment, have important implications for the crisis of democracy. As Thelen calls for political science to consider macro political economy and the institutions of employment, perhaps our field needs to spend more time considering the implications of work and employment on political institutions and politics.

Returning to Thelen's address, she also argues cause and effect, saying that "unions have been the key protagonists in expanding social protections and reducing inequality." Further, "One of the most robust findings in the literature on comparative political economy is that the strength of the organized labor movement is associated with lower inequality (especially low-end inequality) and more generous social protections." Voos points to lower voting rates among those with lower incomes as another explanation for our poor social safety net and labor laws. But I wonder whether there isn't a different cause of poor social protections in the United States and low unionization rates.

And this brings me back to our racial reckoning and to the second book I want to talk about.

This book is *The Sum of Us*, by Heather McGhee, who just happens to be an Ian Haney López student. Where do I even start with this searing and necessary book? McGhee's central argument is that we are all harmed by white racism and especially by the related zero-sum notion held by many white people that if people of color, especially Black people, have more of something, that means they [white people] have less.

According to McGhee, "The zero sum is a story sold by wealthy interests for their own profit, and its persistence requires people desperate enough to buy it." Further, she argues, "The narrative that white people should see the well-being of people of color as a threat to their own is one of the most powerful subterranean stories in America. Until we destroy the idea, opponents of progress can always unearth it and use it to block any collective action that benefits us all."

The book walks us through the history of the zero-sum lie (I'm going to call it a lie in the spirit of the book). Starting with a look at the period of the 1680s and early 1700s, which McGhee calls a deliberate effort to legislate a new hierarchy between poor whites and "unwhite Native and African laborers." McGhee points out—as does Thelen—differences between the United States and Europe:

> When it comes to per capita government spending, the United States is near the bottom of the list of industrial countries, below Latvia and Estonia. … With the exception of about 40 years from the New Deal to the 1970s, the United States has had a weaker commitment to public goods, and to the public good, than every country that possesses anywhere near our wealth.

She discusses how the public goods of that so-called Golden Age in the United States—starting with the New Deal itself (including unions and the NLRA), moving through the implementation of the G.I. Bill (if not its design) to the investment in highways that made suburbs possible—were all predicated on segregated access to these goods.

She tells the story of public swimming pools (and this story told on a radio broadcast or podcast was my original introduction to this book). She starts with the "glory" years of pool building starting in the 1920s, leading eventually to the fight to desegregate public pools. But with desegregation many public pools were privatized or, heartbreakingly, filled in. And the point of the pool story is that, now, *no one* who needed a public pool, including poor and working-class whites, got to have a pool any longer.

The idea that race is at the core of why the United States has fewer public goods, less of a social safety net, fewer protective labor laws, and a lower rate of collective bargaining is not new to me, but McGhee has done an important job in systematically documenting it. It is something our field needs to grapple with, whether through a Critical Race Theory lens, as Lee and Tapia have argued, or some other theoretical perspective. I have long been a big fan of Sandy Jacoby's chapter, in his 1991 edited volume *Masters to Managers*. The chapter is titled "American Exceptionalism Revisited: The Importance of Management." It seems high time for someone to write a new piece with the title "American Exceptionalism Re-Revisited: The Importance of Race."

I have meandered long enough. I will simply conclude by repeating that our world is facing multiple interrelated crises. Our field can be at the center of responding to these multiple crises through our research, teaching, service, and practice. Indeed, it *must be* if our field is to survive and prosper. More important, these crises must be faced if we are to survive and prosper as human beings. But we need to be intentional about it, and we need to expand the range of voices we hear, including those of the workers we study, advocate for, and manage, and those who are most marginalized, like the workers I quote at the beginning of this address.

> Perhaps our field needs to spend more time considering the implications of work and employment on political institutions and politics.

Endnotes

[1]See https://bit.ly/3xK74VN.

[2]See https://nyti.ms/3xL6Ygx.

[3]*Perspectives on Politics* 17, no. 1 (March 2019).

[4]See https://bit.ly/2SjV0LO.

[5]See https://bit.ly/3wSH3ng.

II. LERA Best Posters Session

Speak Up: Employees Voice in Regular Meetings Make Jack a Wise Boy

EUNG IL KIM
Yonsei University

Using data from Korea, I examined how employee voice in regular meetings can result in possible positive outcomes. I found evidence that work–family conflicta and work inequity were decreased, and perceived supervisor support was increased compared with not speaking in meetings. Two critical implications emerge from these findings. First, consistent regular meetings are a critical part of employee voice. Second is that regular meetings with employees allows workers to enhance their voice in and outside of the workplace. These findings lead to the conclusion that employee voice should be encouraged in work meetings.

The Impact of Collective Bargaining and Geographic Location on the Pay Rates of Professional Nurses: A Longitudinal Analysis (1970–1993)

STEPHEN HAVLOVIC
Laurentian University

CHARLES G. SMITH
Otterbein College

NLS data were used to test the impact of collective bargaining and work location on nurse wages (n = 954). Twenty-one percent of the nurses worked in the U.S. South, and 26% worked in rural locations. Nurses working in southern states and rural locations had lower wages, but age had a positive impact on nursing wages. Collective agreements covered 19% of the nurses, but only 2% in the southern states. Results support the positive influence of collective bargaining on wages except for nurses in the South, where those with collective agreement coverage had lower wages. The study concludes with suggestions for future research.

Retooling the Double-Edged Sword: Exploring the Impact of Functional Heterogeneity on Informal Cross-Functional Collaboration

JEONGROCK KIM
TASNEEM OMAR AVA
JALANA ELLIS
DANIELA FEBRES
AMANDA KLAVERT
University of Toronto

Introduction

Cross-functional collaboration refers to cooperation between employees of various functions, integrating their pertinent knowledge and expertise (Denison, Hart, and Kahn 1996; Griffin and Hauser 1996; Olson, Walker, and Ruekert 1995). While scholars have long studied this practice, findings have been equivocal with regard to their outcomes (see for meta-analysis Troy, Hirunyawipada, and Paswan 2008). Where some have found cross-functional teams to positively influence innovation (De Luca and Atuahene-Gima 2007; Song and Parry 1997) and market performance (Leenders and Wierenga 2002), others have found them to have either null or negative impact on innovation (Ancona and Caldwell 1992; Sethi, Smith, and Park 2001) and team performance (Bettenhausen 1991; Dougherty 1992). In response, efforts have been put forth to identify the underlying constructs that elevate and hinder the effectiveness of these teams (Baugh and Graen 1997; Daspit et al. 2013; Denison, Hart, and Kahn 1996; Lovelace, Shapiro, and Weingart 2001; Proehl 1996; Randel and Jaussi 2003; Tekleab et al. 2016).

While they represent meaningful findings in collaboration research, our review of the literature and theoretical development attempts to further the literature by investigating a less explored topic: *informal* cross-functional collaboration. Extant, albeit scant, research investigating informal collaboration define it as spontaneous and opportunistic forms of collaboration (Gutwin et al. 2008). For our investigation, we expand upon this established definition, operationalizing informal cross-functional collaboration (ICFC) as spontaneous and transient collaboration between members of different functions. Previous studies have largely examined trends in the ad hoc collaboration and communication among employees occupying the same space or location (Aldunate, Nussbaum, and Gonzalez 2002; Bellotti and Bly 1996; Kraut et al. 1990; Mejia, Favela, and Moran 2008; Muller et al. 2004). However, no study, to our knowledge, has investigated the influence of diversity of members' functional expertise on informal collaboration. To address this gap in knowledge, we examine a key characteristic of cross-functional collaboration—functional heterogeneity—and the theoretical underpinnings of the ways it can create barriers to ICFC. We also investigate constructs that may moderate this relationship; we review the exacerbating effect of structural and cognitive silos, as well as alleviatory efforts that organizations can implement to offset such hindrances. In sum, we present a theoretical framework that demonstrates the ways in which ICFC becomes more or less effective. Figure 1 illustrates our theoretical model.

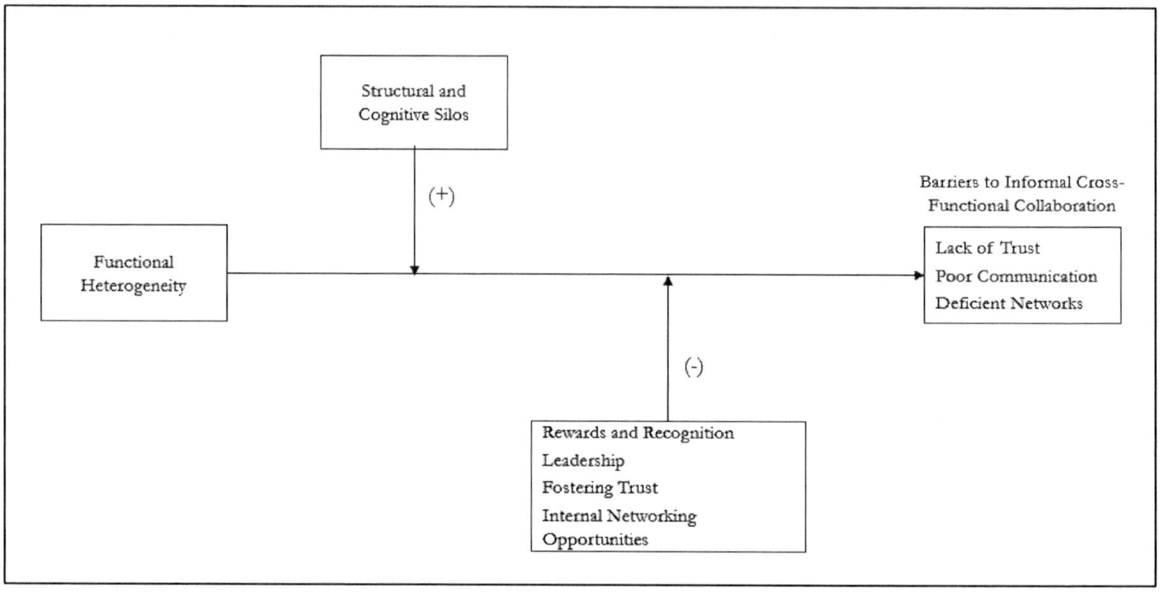

Figure 1. Theoretical model.

Functional Heterogeneity

Scholars have offered numerous descriptions of heterogeneity within the organizational setting. Some have conceptualized it in terms of immutable surface-level characteristics such as race, sex, and age (Earley and Mosakowski 2000; Dahlin, Weingart and Hinds 2005; Kaufmann and Wagner 2017; Kearney and Gebert 2009; Jackson and Joshi 2004), while others have viewed it as divergences in functional expertise and task-orientation (Dougherty 1992; Jackson et al. 1995; Lovelace, Shapiro, and Weingart 2001; Pelled, Eisenhardt, and Xin 1999; Somech 2006). We tailor the current investigation to the latter approach, examining the role of *functional* heterogeneity, which refers to the diversity of pertinent expertise and function rooted within members' respective organizational capacities (Jackson 1992; Somech 2006). Functionally heterogeneous members possess different task-related knowledge, skills, and abilities from one another (Jackson et al. 1995) and thus perform different functions within the organization.

The term "double-edged sword" has been used to describe functional heterogeneity to juxtapose its benefits and pitfalls to organizational outcomes (Bunderson and Sutcliffe 2002; Milliken and Martins 1996). On the one hand, collaboration between functionally heterogeneous members enables broader and deeper information to be accessed (Earley and Mosakowski 2000; Lovelace, Shapiro, and Weingart 2001; Milliken and Martins 1996), which has been shown to produce more innovative approaches in meeting organizational challenges (Bantel and Jackson 1989; Dahlin and Weingart 1996). They are also better suited to implement competitive initiatives (Hambrick, Cho, and Chen 1996) and restructure more quickly by incorporating novel ideas in the decision-making process (Williams, Hoffman, and Lamont 1995), allowing them to more effectively adapt to external threats. On the other hand, functionally heterogeneous members tend to have dissimilar attitudes and are less cohesive, which has negative implications for team performance (Jackson 1992; Swamidass and Aldridge 1996) by hindering clear communication (Bunderson and Sutcliffe 2002; Pelled 1996) and expressions of productive conflict (Bradley et al. 2013). Moreover, negative appraisals and relational conflict between dissimilar members (Hogg and Terry 2000; Tajfel 1982) have been shown to be negatively related to creativity, innovativeness, and team efficiency (Ancona and Caldwell 1992; Lovelace, Shapiro, and Weingart 2001; Yong, Sauer, and Mannix 2014).

These equivocal findings have spurred researchers to model methods to attenuate the costs of cross-functional collaboration such that the benefits offset its costs. Authors have demonstrated that alignment of expectations and goals (Swamidass and Aldridge 1996), exchange of information about members' own functional expertise (Hambrick, Cho, and Chen 1996), and prolonged positive interactions among functionally heterogeneous members to alleviate conflict (Pelled 1996), improve innovativeness and team performance (West 1990; West and Wallace 1991), and ultimately contribute to overcoming the impediments of cross-functional collaboration. Whereas such measures may be effective in traditional cross-functional teams, they may not be suitable for cross-functional collaborations that occur informally. Functional units often have discrete goals and processes (Barki and Pinsonneault 2005) that cannot easily be shifted for short-term collaborations. Improving the management of others' functional knowledge and prolonged positive collaborations requires time investments, which belies the very concept of ICFC. Thus, we argue that functional heterogeneity poses an especially large threat to effective collaboration in this transient form, as it not only maintains the costs of cross-functional work but is also a greater challenge to navigate. Specifically, it is our view that functional heterogeneity is a key precursor to three key barriers to ICFC. In what follows, each of the barriers is discussed.

Functional Heterogeneity and Trust

Despite its negative undertone, some forms of conflict such as expressions of productive conflict (Bradley et al. 2013) and constructive conflict (Cassady 2013) are integral for positive outcomes. Collaboration across functions, however, has more often been associated with harmful forms of conflict (Jones 2006; Lee, Huh, and Reigeluth 2015). Interpersonal trust has been found to curb the costs and risks associated with such collaboration by enabling information sharing and task coordination (Temby et al. 2017; Terman, Feiock, and Youm 2020). In other words, absence of trust prevents effective collaboration due to a fear of engaging in constructive conflict (Cassady 2013).

The adverse effects of an absence or low levels of trust become particularly profound in cross-functional settings, where the mere presence of functional diversity can erode—and prevent the establishment of—trust and positive intergroup relations (Garrison et al. 2010; Sy and Côté 2004; Zolin et al. 2004). The relationship between structural factors of diversity and trust is complex: while some level of diversity is better than no diversity, too much diversity can threaten the development of interpersonal trust, particularly among functionally diverse members (Dayan and Di Benedetto 2010; Knouse and Dansby 1999), as people tend to trust those with whom they share similar, rather than different, backgrounds (Brewer 2008; Foddy, Platow and Yamagishi 2009; Xin, Xin, and Lin 2016). Where there exist benefits to be reaped from cross-functional collaboration, low levels of trust make doing so difficult. With low affect-based trust, functionally diverse members are often reluctant to engage in knowledge sharing, which in turn diminishes innovation (Cheung et al. 2016). Employees' apprehension may stem from the perception that shared knowledge may be used to make private gains and outperform others (Ghobadi and D'Ambra 2012; Riege 2005). Such a climate of competition can reduce out-group liking (Montoya and Pittinsky 2011), which in turn lessens members' willingness to seek help from their functionally heterogeneous counterparts (Van Leeuwen, Täuber and Sassenberg 2011) and increases knowledge hoarding (Garrison et al. 2010).

Trust can be built through long-term professional and interpersonal ties (Temby et al. 2017; Wulf and Butel 2017). Once a reputation of trustworthiness has been built through this process, collaboration becomes less dependent on rigid organizational role relationships (Thomson, Perry, and Miller 2009); that is, establishing trust takes time and requires repeated interaction between collaborators (Buvik and Rolfsen 2015; Thomson, Perry, and Miller 2009; Webber 2008) which, by the nature of ICFC, is not facilitated. Thus, functional heterogeneity is particularly detrimental to developing trust in the context of informal collaboration.

Proposition 1. Functional heterogeneity will be negatively related to trust between informal collaborators.

Functional Heterogeneity and Communication

The physical layout of the workplace and the location of units within a building are likely to impact communicative patterns among employees. Closer physical proximity increases the frequency of face-to-face interactions between functionally diverse members, which can elevate the quality of communication to levels comparable to those between members of the same functional unit (Gray, Siemsen, and Vasudeva 2015; Pinto, Pinto, and Prescott 1993; Van den Bulte and Moenaert 1998). However, organizations tend to place its members in proximity only to those who serve similar functions (Hoegl and Proserpio 2004), granting fewer opportunities to interact with functionally diverse colleagues.

In addition to this structural barrier, forms of communication unique to each function may also hinder effective collaboration across functions. For example, function-specific jargon that does not invoke the same definitions in alternative contexts raises difficulties for members outside of the unit, making the establishment of common ground difficult (Bechky 2003). Further, each function may use distinct forms of communication, spanning from synchronous (in-person meetings) to asynchronous (emails) (Patrashkova and McComb 2004). Selecting the appropriate mode of communication is crucial for the direction of information flow (Dussart, van Oortmerssen, and Albronda 2021); where one function may utilize one system, another may not, producing logistical difficulties in communication.

Lastly, out-group biases often form against those who occupy different functions (Ancona and Caldwell 1992; Cantner, Goethner, and Stuetzer 2010; Hogg and Terry 2000), and the ensuant dynamics of secrecy and hostility (Holland, Gaston, and Gomes 2000) make sparse the willingness to share information across functions. Indeed, poor quality of communication is cited as a large contributor to employees' negative experiences when collaborating with functionally heterogeneous others (Kaye and Cook 2015). Thus, functional heterogeneity's propensity to engender such cognitive, structural, and logistical barriers compromise members' ability to effectively communicate with members of different units. Given that the realization of ICFC's potential to produce positive outcomes is heavily contingent on a high frequency of high-quality interactions between functionally heterogeneous members (Temby et al. 2017; Whittaker, Frohlich, and Daly-Jones 1994), these conditions represent especially salient barriers.

Proposition 2. Functional heterogeneity will be negatively related to the frequency and quality of communication between cross-functional collaborators.

Functional Heterogeneity and Networks

Although diverse knowledge and skills enable functionally heterogeneous members to contribute to the organization in unique ways, the double-edged nature of functional heterogeneity can also lead to a lack of cohesion among these dissimilar members (Swamidass and Aldridge 1996). Functional divergences may hinder swift communication and exchange of information among such members (Bunderson and Sutcliffe 2002; Pelled 1996), which are two critical factors in the process of network formation (Bala and Goyal 2000). Further, collaboration among functionally heterogeneous members can evoke conflict on both relational and task-related dimensions (Curşeu and Schruijer 2010; Hogg and Terry 2000) while decreasing the likelihood of informal interactions (Bento and Garotti 2019), hindering the formation of close and trustful networks across the organization.

Moreover, this difficulty for functionally heterogeneous members to form ties hinders their ability to build relational capital, or the quality of members' relationships with others across the organization (Moran 2005), which is strongly related to several key enablers of cross-functional collaboration including trust (Castelfranchi, Falcone and Marzo 2006; Gittell 2002; Kale, Singh, and Perlmutter 2000), cohesion (Granovetter 1992; Moran 2005), goal alignment (Adler and Kwon 2002; McGrath 2001; Mom et al. 2015), and knowledge sharing (Aisyah, Sukoco, and Anshori 2019; Maurer, Bartsch, and Ebers 2011; Wu, Liao, and Dai 2015). Relational capital has also been identified as an antecedent of members' orientation toward innovativeness (Kijkuit and Van den Ende 2010; Moran 2005), which may partially explain previous findings of functionally diverse teams' innovativeness (e.g., Bantel and Jackson 1989; Dahlin and Weingart 1996). Such conditions afforded through relational capital are critical contributors in enabling effective cross-functional collaboration (Thomson et al. 2009; Tsai and Hsu 2014; West 1990). However, functional heterogeneity gives rise to conditions that limit the formation of networks and professional ties across functions, hindering members' ability to build relational capital within the organization.

Networks and relational capital are particularly important when engaging in ICFC. As this type of collaboration is not formally mandated and occurs spontaneously out of a member's own volition, the ability to leverage their networks with others within the organization is a key necessity. In other words, members' ability to effectively collaborate with those of other functions is commensurate with the size of their personal network (Lee and Shin 2017), which is likely negatively related to functional heterogeneity. In addition to the extent of one's network, the quality of those professional ties is also important in the sharing of knowledge across organizational units. However, high levels of trust and interaction between collaborators have been found to be crucial in developing relational capital (Castelfranchi, Falcone, and Marzo 2006; Liu, Ghauri, and Sinkovics 2010; Zornoza, Orengo, and Peñarroja 2009), which are deficient in the process ICFC. On the basis of such findings, it can be seen that functional heterogeneity also poses a particularly large threat to ICFC by limiting members' ability to leverage the extent of their network and the quality of those ties.

Proposition 3. Functional heterogeneity will be negatively related to the extent of members' networks and relational capital.

The Moderating Effect of Structural and Cognitive Silos

According to Tajfel's (1982) social categorization processes, individuals categorize themselves into groups with whom they share key characteristics and associate frequently. They identify with those groups, striving to maintain positive appraisals of the group while developing negative ones of those whose identity is dissimilar from their own. This process extends to the organizational setting, wherein members not only form tighter networks with those of similar in-group prototypes, but they also come to like members of different groups less and tend to maximize the dimensions that characterize their differences (Hogg and Terry 2000; Van Knippenberg and Schippers 2007). Owing to their importance for successful performance, job- and task-related dimensions such as functional knowledge are particularly likely to lead members to make categorizations according to these attributes (Ancona and Caldwell 1992; Pelled 1996; Webber and Donahue 2001). In other words, members are likely to identify and categorize themselves into groups among functionally homogeneous members (the "in-group") while developing negative biases toward functionally heterogeneous others (the "out-group").

This process may occur as a result of negative experiences involved in collaborating with members of the functional out-group. Differences in functional capacity foster divergent perceptions and approaches to tasks (Waller, Huber, and Glick 1995); this incongruence may provoke task-related conflict between collaborating members (Pelled, Eisenhardt, and Xin 1999), which can incite relational conflict (De Dreu and Weingart 2003). Indeed, functional heterogeneity has been found to be negatively related to collaborator trustworthiness and satisfaction of the collaboration (Tuer 2013). Such conflict may lay the groundwork for limited information sharing channels between functions. Jackson and colleagues (1995) posit that because of biases, members are more likely to seek and offer information between functionally homogeneous members than heterogeneous ones. They further argue that a consolidation of social dynamics occurs such that networks are formed solely among functionally homogeneous members. The over-identification and preference of functionally homogeneous members is conceptually analogous to a silo mentality, a cognitive phenomenon that serves to inhibit cross-functional collaboration (Alves and Meneses 2018).

Functional silos refer to groups of employees entrenched within their functions and acting autonomously from other units (Vatanpour, Khorramnia, and Forutan 2013). When a silo mentality permeates across functional silos, units develop unique internal cultures and develop barriers to communication and information sharing, which hinders collaborative work processes (AME 1988; Fenwick, Seville, and Brunsdon 2009). At its worst, a silo mentality may engender dynamics of hostility between functions (Barmyer and Sachseneder 2013), wherein each unit competes over organizational resources and prioritizes their own goals over those of other units and even that of the organization (Fenwick, Sevile, and Brunsdon 2009; Schütz and Bloch 2006; Stone 2004; Sy and Côté 2004). Such structural and cognitive impediments diminish cross-functional collaboration's potential to produce positive outcomes. Taken in the context of our investigation, we argue that silos can exacerbate the barriers to trust, communication, and network formation already presented by functional heterogeneity.

Silos' Impact on Trust, Communication, Networks

Functional heterogeneity's role as a barrier of collaboration becomes more pronounced in the presence of structural and cognitive silos, as they can be detrimental to the development of interpersonal trust between members of separate units (Vatanpour, Khorramnia, and Forutan 2013). Studies have consistently demonstrated the prominence of trust and collaborative behaviors among members of the same social category (Brewer 2008; Montoya and Pittinsky 2011; Xin, Xin, and Lin 2016), even when in-group members are characterized by negative stereotypes (Foddy, Platow, and Yamagishi 2009). This form of depersonalized trust, in which trust is generalized and given solely based on shared group membership rather than individualized knowledge (Balliet, Wu, and De Dreu 2014; Brewer 1996; Kramer 2010), can prevent productive amalgamation of diverse perspectives and therefore thwart effective collaboration across functions.

Functional and cognitive silos most often manifest in the form of communication barriers, hindering effective work processes (Fenwick, Seville, and Brunsdon 2009; Sessoms 2017). Indeed, difficulties in establishing shared vernacular and divergences in preferred modes of communication do not facilitate effortless communication. Meager levels of trust that permeate functional silos may diminish members' willingness to bridge this gap, which can contribute to a lack of awareness of other units' workflows (Kaye and Cook 2015; Terman, Feiock, and Youm 2020), poor knowledge and information sharing (Bureš 2003; Riege 2005), cultural tension (Bundred 2006), and even an environment of competition (Alves and Meneses 2018). Given that scarce communication begets isolation and hampers functional integration (Vatanpour, Khorramia, and Forutan 2013), the establishment of silos likely precipitate barriers to communication to a greater extent than the ones already posed by functional heterogeneity.

Networks are social structures comprising the movement of new and old members participating to generate information flows (Bento, Tagliabue, and Lorenzo 2020). Clusters of networks throughout an organization are significant for enabling the proliferation of information among employees (Centola 2018; Guilbeault, Becker, and Centola 2018). Circulating complex information, however, requires tightly knit networks as well as a medium to share the information (Choi, Kang, and Lee 2008; Widén-Wulff and Ginman 2004). Due to the paucity of trust and negative appraisals attached to functionally heterogeneous members, close interpersonal networks are unlikely to be formed. Moreover, members often lack access to an appropriate medium in which to engage in informal information sharing, such as shared physical workspaces. In other words, silos isolate employees by restricting them from developing deeper relationships with colleagues across functions (de Waal et al. 2019) by establishing cognitive and structural impediments to do so. Thus, functional silos and silo mentality further inhibits employees' ability to form networks across the organization and impedes their ability to leverage their relational capital.

Proposition 4. Structural and cognitive silos will moderate the positive relationship between functional heterogeneity and barriers to ICFC such that when structural and cognitive silos are established, the relationship becomes stronger.

The Moderating Effect of Organizational Efforts

Rewarding Informal Collaboration

Like many other behaviors in the organizational setting, informal collaboration may be fostered through rewards. Be it in intrinsic forms like recognition or extrinsic like financial incentives, rewards can help drive informal collaboration (Peterson and Luthans 2006; Wei and Yazdanifard 2014). For them to be effective, rewards must be clear, well-communicated, and widely understood by employees across all levels of the organization (Ghobadi, Campbell and Clegg 2017). To encourage sustained collaboration, employees must also consistently expect to be rewarded (Engelsberger et al. 2021) and have such behaviors appreciated (Bishop 1999) for demonstrating collaborative behaviors across functions.

Given that tangible rewards are inextricably tied to performance appraisals, systems of evaluating employees' behaviors should be designed to promote ICFC. This can be accomplished by assessing employees' cooperation outside of their core functional unit (Ellinger, Keller and Ellinger 2000; Lee and Shin 2017) and evaluating them not only on individual performance but also on joint performance (Ghobadi, Campbell, and Clegg 2017; Lee and Shin 2017). Indeed, Bartol and Srivastava (2002) remark that such rewards granted on a communal basis can foster knowledge sharing across functions. They further posit that employees who receive tangible rewards for engaging in collaborative behaviors can develop a sense of organizational ownership and commitment.

Intrinsic rewards can be an equally potent enabler of ICFC; reinforcing collaborative behavior by recognizing employees who engage in positive activities like sharing tacit knowledge to help others achieve their goals can create a sense of obligation to continue engaging in this behavior (Song 2009). In fact, praise and public recognition have been found to facilitate knowledge sharing to a greater extent than material rewards (Choi, Kang, and Lee 2008). As such, employee recognition programs can be utilized to develop relationships within the organization by promoting the attainment of common goals through collaboration (Dhanpat 2014). In sum, consistent reward and recognition for engaging in ICFC can transform the act of collaborating across functions into a widely accepted and normative behavior, for which employees can expect reciprocity from others and the organization.

Proposition 5. Organizational policies designed to reward and recognize ICFC will moderate the positive relationship between functional heterogeneity and barriers to ICFC such that when these policies are implemented and widely utilized, the relationship becomes weaker.

Fostering Trust

Insofar as rewards and recognition programs contribute to members' willingness and frequency of engaging in collaborative behaviors, the impact is indirect, and its importance is preceded by the role of trust between functionally heterogeneous members. In facilitating knowledge sharing through informal means, trust is a necessary precondition (Abrams et al. 2003; Bartol and Srivastava 2002). Two dimensions of trust related to knowledge sharing include benevolence and competence trust (Abrams et al. 2003). As an affective construct, benevolence trust is important for knowledge sharing in informal networks, as request for help inherently carries an undertone of vulnerability. Likewise, as a cognitive aspect of trust, competence trust is required to establish a belief that the shared knowledge originates from a source who possesses the requisite professional expertise and skills to address the issue (Abrams et al. 2003; Choi, Kang, and Lee 2008). With greater levels of competence trust, functionally heterogeneous members may engage in productive task conflict, openly sharing diverse perspectives to produce more positive team outcomes (Olson, Parayitam, and Bao 2007). Thus, organizations that employ strategies to foster trust between functionally diverse employees are likely to be better suited to negate the trust barriers that arise from functional heterogeneity.

Proposition 6. Organizational efforts toward fostering trust between functionally heterogeneous members will moderate the positive relationship between functional heterogeneity and barriers to ICFC such that when these efforts are put forth, the relationship becomes weaker.

Leadership

A major challenge of collaboration between functionally heterogeneous members is the inherent lack of team cohesion and trust (Ghobadi and D'Ambra 2012; Sy and Côté 2004). Webber (2002) suggests that leader behavior oriented toward facilitating collaboration, including setting clear expectations, fostering positive relationships between members, and articulating shared commitment toward a common goal can mediate such challenges. Indeed, empirical support has been found for enhanced cohesion and trust, and reduced conflict among such members with leaders' orientation toward establishing and maintaining positive working relationships (Curşeu 2011; Koohang, Paliszkiewicz, and Goluchowski 2017). Similarly, leader support and empowerment have been found to indirectly impact goal congruence among members by improving morale, which in turn motivate employees to align their goals with others (Coote, Price, and Ackfeldt 2004).

Two styles of leadership may be particularly effective in encouraging collaborative behaviors across functional units. Participative leaders facilitate team reflection, a process by which the diverse knowledge, skills, and abilities of members become actualized to the team's benefit (West 2002) by enhancing communication and reducing ambiguity (Latham, Winters, and Locke 1994; Somech 2006). Transformational leadership, in addition to facilitating goal alignment (Bass and Riggio 2006), has been found to facilitate team identification among members of diverse educational backgrounds (Kearney and Gebert 2009). Given that members who identify with one another are more likely to engage in sharing their function-specific skills and knowledge (Shin and Zhou 2007), transformational leaders enable functionally heterogeneous members to collaborate with more ease and contribute to team outcomes in more robust ways. Leaders' orientation and behaviors toward facilitating collaboration, therefore, is a crucial component of alleviating the challenges involved in collaborating across functions.

Proposition 7. Leadership oriented toward facilitating informal collaboration will moderate the positive relationship between functional heterogeneity and barriers to ICFC such that when this style of leadership is prominent, the relationship becomes weaker.

Internal Networking Opportunities

Organizations that provide ample opportunities for social interaction and the formation of networks see higher levels of trust (Ahuja, Soda, and Zaheer 2012; Wu, Liao, and Dai 2009). Given that trust is closely associated with knowledge sharing behaviors (Abrams et al. 2003; Bartol and Srivastava 2002), providing employees with the medium to build extensive networks across the organization may be a key enabler of ICFC. In the context of our investigation, whereas social categorization processes create relational distance among functionally heterogeneous members, networking opportunities represent avenues by which to form connections based on characteristics distinct from members' functional capacities. Internal networking opportunities can be in the form of mentoring programs, after-work activities, workshops, conferences, cross-functional teams, relationship programs, co-locational collaborative spaces, use of facilitators, and joint training (Cousins, Lawson, and Squire 2008; Gupta and Govindarajan 2000; Suprapto et al. 2015; Turkulainen and Ketokivi 2012). Given that knowledge sharing through informal means, as opposed to rigid knowledge sharing structures, may be more favored among employees (Jewels, Underwood, and de Pablos Heredero 2003), implementing programs for employees to form informal networks and build relational capital across the organization may attenuate the networks barrier presented by functional heterogeneity.

Proposition 8. Programs designed to aid in the formation of networks between members across functions will moderate the positive relationship between functional heterogeneity and barriers to ICFC, such that when these programs are implemented and widely utilized, the relationship becomes weaker.

Discussion

To investigate the process by which functional heterogeneity produces barriers to ICFC, we review the extant literature to arrive at a conceptual model. Our model proposes that functional heterogeneity will have negative implications on dimensions related to trust, communication, and network formational utilization in the context of collaborating with members outside of their functional units. It also proposes two potential moderators to this relationship: cognitive and structural silos, and alleviatory organizational efforts. Our review of the theoretical and empirical findings in relevant literature yields the propositions that silos will strengthen the degree to which functional heterogeneity hinders ICFC, while alleviatory organizational efforts attenuate such negative implications.

Theoretical Contributions

The current review has several implications for the collaboration and diversity literature. First, our review is among the first to examine informal collaboration in a cross-functional context. Our investigation of ICFC represents a divergence in the research approaches that several scholars have taken; whereas cross-functional collaboration has been thoroughly investigated by numerous scholars (Ancona and Caldwell 1992; Denison, Hart, and Kahn 1996; Lovelace, Shapiro, and Weingart 2001; Troy, Hirunyawipada, and Paswan 2008), to the best of our knowledge, our review is the first to examine a specific type of cross-functional collaboration.

Second, we propose that despite being an inherent characteristic of cross-functional collaboration, functional heterogeneity can be an impediment in informal contexts. Whereas the impact of member diversity on task-related dimensions have been examined by a number of scholars (Cannella, Park, and Lee 2008; Kearney, Gebert, and Voelpel 2009; Phillips and Loyd 2006; Randel and Jaussi 2003; Stewart and Johnson 2009), its impact on informal collaboration remains, as yet, an uninvestigated area of the literature. Our investigation into the impact of functional diversity on informal collaboration is thus a meaningful departure from previous lines of research. Moreover, by addressing the dearth of research in the literature surrounding this type of collaboration, our review furthers the literature and initiates future theoretical and empirical research on this novel topic.

Finally, we propose two moderators, cognitive and structural silos, and alleviatory organizational efforts, and examine their impact on the initial relationship between functional heterogeneity and barriers to ICFC. We argue that silos raise structural and cognitive challenges that can exacerbate the issues presented by functional heterogeneity, while organizational efforts to alleviate such problems can help organizations actualize the potential of ICFC. In sum, the theoretical model we develop presents an integrated view of the numerous constructs and the interplay therein to inform the process by which the effectiveness of a firm's ICFC practices may be determined.

Practical Implications

By separating the conceptual underlay between different forms of cross-functional collaboration, our research offers a novel perspective with which to examine employees' regular work processes. The delineation between formal and informal cross-functional collaboration is an important one, as not all forms of collaboration are formally mandated to create cross-functional teams. Further, informal collaborations serve a useful purpose different from those of formal ones; they are more appropriate to address unanticipated problems, especially in more complex projects (Kraut and Streeter 1995), and occur more frequently due to the ease with which they are initiated (Gutwin et al. 2008; Kraut et al. 1990). Given that organizations are turning to integrating the knowledge of functionally heterogeneous members in order to face the increasingly dynamic and competitive markets (Dinca and Voinescu 2012; Lovelace, Shapiro, and Weingart 2001), a deeper understanding of the ways in which this process takes place informally may have immense practical implications for individual employees and organizations alike.

In addition, our review into the organizational efforts that can alleviate the barriers presented by functional heterogeneity may help shape organizations' human resources management practices. We predict that our research will have significant implications for large corporations that operate a wide range of functions; while they may be more susceptible to experience the pitfalls of functional heterogeneity and silos, they may have greater resources to implement programs to attenuate them in the form of modified rewards and recognition policies, tailored leadership training, and networking opportunities. Thus, identifying the difficulties of ICFC and putting forth initiatives to alleviate them may be of particular use for larger corporations.

Limitations and Future Research

While our research and proposed model adds to the limited literature on informal collaboration among functionally diverse employees, a major limitation of our current research is the lack of empirical data. Moreover, our research examines solely task-related (i.e., functional) diversity, and does not consider other types of diversity. Van Knippenberg and Schippers (2007) have suggested the need for a more integrative conceptualization of diversity, in which the various dimensions of diversity are considered and examined simultaneously, rather than attempting to isolate the costs and benefits of any one specific type. As such, we call for future research to test our model and additionally explore the relationship between the various types of diversity and informal collaboration. As research in this field remains relatively scarce, further research is required to develop greater theoretical understanding with which to empirically test functional heterogeneity and its multidimensional impact on ICFC. Another important direction for future research concerns the role of communication media and technology and its influence on collaboration. Virtual collaboration has never been more relevant than today as the pandemic pushes us to become even more dependent on technology. However, such forms of communication may present its own unique set of difficulties, especially in the context of informal collaboration. Thus, investigation into the costs and benefits of virtual communication on both formal and informal forms of collaboration will be valuable.

References

Abrams, Lisa C., Rob Cross, Eric Lesser, and Daniel Z. Levin. 2003. "Nurturing interpersonal trust in knowledge-sharing networks." *Academy of Management Perspectives,* Vol. 17, no. 4, pp. 64-77.

Adler, Paul S., and Seok-Woo Kwon. 2002. "Social capital: Prospects for a new concept." *Academy of Management Review,* Vol. 27, no. 1, pp. 17-40.

Ahuja, Gautam, Giuseppe Soda, and Akbar Zaheer. 2012. "The genesis and dynamics of organizational networks." *Organization Science,* Vol. 23, no. 2, pp. 434-448.

Aisyah, Ratri Amelia, Badri Munir Sukoco, and Muslich Anshori. 2019. "The effect of relational capital on performance: knowledge sharing as mediation variables in supplier and buyer relation." *International Journal of Logistics Systems and Management,* Vol. 34, no. 2, pp. 211-232.

Aldunate, Roberto, Miguel Nussbaum, and Roberto Gonzalez. 2002. "An agent-based middleware for supporting spontaneous collaboration among co-located, mobile, and not necessarily known people." In *Workshop on Ad-hoc Communications and Collaboration in Ubiquitous Computing Environments, ACM CSCW.*

Alves, João, and Raquel Meneses. 2018. "Silo mentality in healthcare services." *11th Annual Conference of the EuroMed Academy of Business* (Valletta, Sep. 12-14, 2018). Research Advancements in National and Global Business Theory and Practice, pp. 65-79.

AME Study Group on Functional Organization. 1988. "Organizational renewal: Tearing down the functional silos."

Ancona, Deborah Gladstein, and David F. Caldwell. 1992. "Demography and design: Predictors of new product team performance." *Organization Science,* Vol. 3, no. 33, pp. 321-341.

Bala, Venkatesh, and Sanjeev Goyal. 2000. "A noncooperative model of network formation." *Econometrica,* Vol. 68, no. 5, pp. 1181-1229.

Balliet, Daniel, Junhui Wu, and Carsten KW De Dreu. 2014. "Ingroup favoritism in cooperation: a meta-analysis." *Psychological Bulletin,* Vol. 140, no. 6, pp. 1556-1581.

Bantel, Karen A., and Susan E. Jackson. 1989. "Top management and innovations in banking: Does the composition of the top team make a difference?" *Strategic Management Journal,* Vol. 10, no. S1, pp. 107-124.

Barki, Henri, and Alain Pinsonneault. 2005. "A model of organizational integration, implementation effort, and performance." *Organization Science,* Vol. 16, no. 2, pp. 165-179.

Barmyer, C., and C. Sachseneder. 2013. "From silo mentality to 'angularity' of the departments. The contribution of departmental cultures to diversity in companies." *European Academy of Management Conference* (Istanbul, June 26-29, 2013).

Bartol, Kathryn M., and Abhishek Srivastava. 2002. "Encouraging knowledge sharing: The role of organizational reward systems." *Journal of Leadership & Organizational Studies,* Vol. 9, no. 1, pp. 64-76.

Baugh, S. Gayle, and George B. Graen. 1997. "Effects of team gender and racial composition on perceptions of team performance in cross-functional teams." *Group & Organization Management,* Vol. 22, no. 3, pp. 366-383.

Bass, Bernard M., and Ronald E. Riggio. 2006. "Transformational leadership." New York: Psychology Press.

Bechky, Beth A. 2003. " Sharing Meaning Across Occupational Communities: The Transformation of Understanding on a Production Floor." *Journal of Organization Science,* Vol. 14, no. 3, pp. 312-330.

Bellotti, Victoria, and Sara Bly. 1996. "Walking away from the desktop computer: distributed collaboration and mobility in a product design team." *Proceedings Of The 1996 ACM Conference on Computer Supported Cooperative Work* (Boston, November, 1996), New York, NY: Association For Computing Machinery, Pp. 209-218.

Bento, Fabio., and Luciano Garotti. 2019. "Resilience beyond Formal Structures: A Network Perspective towards the Challenges of an Aging Workforce in the Oil and Gas Industry" *Journal of Open Innovation: Technology, Market, and Complexity,* Vol. 5, no. 1, pp. 15.

Bento, Fabio., Marco Tagliabue, and Flora Lorenzo. 2020. "Organizational Silos: A Scoping Review Informed by a Behavioral Perspective on Systems and Networks" *Societies,* Vol. 10, no. 3, pp. 56.

Bettenhausen, Kenneth L. 1991. "Five years of groups research: What we have learned and what needs to be addressed." *Journal of Management,* Vol. 17, no. 2, pp. 345-381.

Bishop, Suzanne. K. 1999. "Cross-functional project teams in functionally aligned organizations." *Project Management Journal,* Vol. 30, no. 3, pp. 6-12.

Bradley, Bret H., Anthony C. Klotz, Bennett E. Postlethwaite, and Kenneth G. Brown. 2013. "Ready to rumble: How team personality composition and task conflict interact to improve performance." *Journal of Applied Psychology,* Vol. 98, no. 2, pp. 385-392.

Brewer, Marilynn B. 2008. "Depersonalized trust and ingroup cooperation." In J. I. Krueger (Ed.), *Modern Pioneers in Psychological Science: An APS-Psychology Press Series. Rationality and Social Responsibility: Essays in Honor of Robyn Mason Dawes,* pp. 215-232. Psychology Press.

Bunderson, J. Stuart, and Kathleen M. Sutcliffe. 2002. "Comparing alternative conceptualizations of functional diversity in management teams: Process and performance effects." *Academy of Management Journal,* Vol. 45, no. 5, pp. 875-893.

Bundred, Steve. 2006. "Solutions to silos: Joining up knowledge." *Public Money and Management, Vol.* 26, no. 2, pp. 125-130.

Bureš, Vladimír. 2003. "Cultural barriers in knowledge sharing." *E + M Economics and Management, Vol.* 6, special issue, pp. 57-62.

Buvik, Marte Pettersen, and Monica Rolfsen. 2015. "Prior ties and trust development in project teams–A case study from the construction industry." *International Journal of Project Management,* Vol. 33, no. 7, pp. 1484-1494.

Cannella Jr, Albert A., Jong-Hun Park, and Ho-Uk Lee. 2008. "Top management team functional background diversity and firm performance: Examining the roles of team member colocation and environmental uncertainty." *Academy of Management Journal,* Vol. 51, no. 4, pp. 768-784.

Cantner, Uwe, Maximilian Goethner, and Michael Stuetzer. 2010. "Disentangling the effects of new venture team functional heterogeneity on new venture performance." *JENA Economic Research Papers,* no. 2010, 029.

Cassady, Sandra. 2013. "The Linda Crane lecture from silos to bridges: preparing effective teams for a better delivery system." *Cardiopulmonary Physical Therapy Journal,* Vol. 24, no. 2, pp. 5-11.

Castelfranchi, Cristiano, Rino Falcone, and Francesca Marzo. 2006. "Being trusted in a social network: Trust as relational capital." *Proceedings of the 4th Annual International Conference on Trust Management* (Pisa, May 17-19, 2006). Berlin, Heidelberg: Springer, pp. 19-32.

Centola, Damon. 2018. *How Behavior Spreads: The Science of Complex Contagions.* New Jersey: Princeton University Press.

Curşeu, Petru Lucian, and Sandra GL Schruijer. 2010. "Does conflict shatter trust or does trust obliterate conflict? Revisiting the relationships between team diversity, conflict, and trust." *Group Dynamics: Theory, Research, and Practice,* Vol. 14, no. 1, pp. 66-79.

Cheung, Siu Yin, Yaping Gong, Mo Wang, Le Zhou, and Junqi Shi. 2016. "When and how does functional diversity influence team innovation? The mediating role of knowledge sharing and the moderation role of affect-based trust in a team." *Human Relations,* Vol. 69, no. 7, pp. 1507-1531.

Choi, Sue Young, Young Sik Kang, and Heeseok Lee. 2008. "The effects of socio-technical enablers on knowledge sharing: an exploratory examination." *Journal of Information Science,* Vol. 34, no. 5, pp. 742-754.

Coote, Leonard V., Evan Price, and Anna-Lena Ackfeldt. 2004. "An investigation into the antecedents of goal congruence in retail-service settings." *Journal of Services Marketing, Vol.* 18, no. 7, pp. 547-559.

Cousins, Paul D., Benn Lawson, and Brian Squire. 2008. "Performance measurement in strategic buyer-supplier relationships." *International Journal of Operations & Production Management, Vol.* 28, no. 3, pp. 238-258.

Curşeu, Petru L. 2011. "Intra-group conflict and teamwork quality: The moderating role of leadership styles." *Administrative Sciences,* Vol. 1, no. 1, pp. 3-13.

Dahlin, Kristina B., and Laurie. R. Weingart. 1996. "Absorptive capacity—A link between group diversity and group performance." *Annual Meeting of the Academy of Management* (Cincinnati,1996).

Dahlin, Kristina B., Laurie R. Weingart, and Pamela J. Hinds. 2005. "Team diversity and information use." *Academy of Management Journal,* Vol. 48, no. 6, pp. 1107-1123.

Daspit, Josh, C. Justice Tillman, Nancy G. Boyd, and Victoria McKee. 2013. "Cross-functional team effectiveness: An examination of internal team environment, shared leadership, and cohesion influences." *Team Performance Management: An International Journal,* Vol. 19, no. 1, pp. 34-56.

Dayan, Mumin, and C. Anthony Di Benedetto. 2010. "The impact of structural and contextual factors on trust formation in product development teams." *Industrial Marketing Management,* Vol. 39, no. 4, pp. 691-703.

De Dreu, Carsten KW, and Laurie R. Weingart. 2003. "Task versus relationship conflict, team performance, and team member satisfaction: a meta-analysis." *Journal of Applied Psychology,* Vol. 88, no. 4, pp. 741-749.

De Luca, Luigi M., and Kwaku Atuahene-Gima. 2007. "Market knowledge dimensions and cross-functional collaboration: Examining the different routes to product innovation performance." *Journal of Marketing,* Vol. 71, no. 1, pp. 95-112.

Denison, Daniel R., Stuart L. Hart, and Joel A. Kahn. 1996. "From chimneys to cross-functional teams: Developing and validating a diagnostic model." *Academy of Management Journal, Vol.* 39, no. 4, pp. 1005-1023.

de Waal, André., Weaver, Michael., Day, Tammy., and van der Heijden, Beatrice. 2019. "Silo-Busting: Overcoming the Greatest Threat to Organizational Performance." *Sustainability*, Vol. 11, no. 23, pp. 6860.

Dhanpat, Nelesh. 2014. "An Investigation: Should Organizations Move towards Team-Based Recognition?" *Mediterranean Journal of Social Sciences*, Vol. 5, no. 23, pp. 2359-2359.

Dinca, Laura, and Carmen Voinescu. 2012. "Cross-functional teams and their role in increasing competitiveness of the organizational partnerships." *EIRP Proceedings, Vol.* 7, no. 1, pp. 453-459.

Dougherty, Deborah. 1992. "Interpretive barriers to successful product innovation in large firms." *Organization Science*, Vol. 3, no. 2, pp. 179-202.

Dussart, Pascal, Lise A. Van Oortmerssen, and Bé Albronda. 2021. "Perspectives on knowledge integration in cross-functional teams in information systems development." *Team Performance Management: An International Journal.*

Earley, Christopher P., and Elaine Mosakowski. 2000. "Creating hybrid team cultures: An empirical test of transnational team functioning." *Academy of Management Journal*, Vol. 43, no. 1, pp. 26-49.

Ellinger, Alexander E., Scott B. Keller, and Andrea D. Ellinger. 2000. "Developing interdepartmental integration: An evaluation of three strategic approaches for performance improvement." *Performance Improvement Quarterly,* Vol. 13, no. 3, pp. 41-59.

Engelsberger. A , J. Cavanagh, T. Bartram, and B. Halvorsen. 2021. "Multicultural skills in open innovation: relational leadership enabling knowledge sourcing and sharing." *Personnel Review*, Vol. ahead-of-print, no. ahead of print.

Fenwick, Tony, Erica Seville, and Dave Brunsdon. 2009. "Reducing the impact of organizational silos on resilience." *Resilient Organizations Research Programme.*

Foddy, Margaret, Michael J. Platow, and Toshio Yamagishi. 2009. "Group-based trust in strangers: The role of stereotypes and expectations." *Psychological Science,* Vol. 20, no. 4, pp. 419-422.

Garrison, Gary, Robin L. Wakefield, Xiaobo Xu, and Sang Hyun Kim. 2010. "Globally distributed teams: The effect of diversity on trust, cohesion and individual performance." *ACM SIGMIS Database: The Database for Advances in Information Systems,* Vol. 41, no. 3, pp. 27-48.

Ghobadi, Shahla, John Campbell, and Stewart Clegg. 2017. "Pair programming teams and high-quality knowledge sharing: A comparative study of cooperative reward structures." *Information Systems Frontiers,* Vol. 19, no. 2, pp. 397-409.

Ghobadi, Shahla, and John D'Ambra. 2012. "Knowledge sharing in cross-functional teams: A competitive model." *Journal of Knowledge Management,* Vol. 16, no. 2, pp. 285-301.

Gittell, Hoffer. 2002. "Coordinating mechanisms in care provider groups: Relational coordination as a mediator and input uncertainty as a moderator of performance effects." *Management Science,* Vol. 48, no. 11, pp. 1408-1426.

Granovetter, Mark. 1992. "Problems of explanation in economic sociology." In Nitin Nohria and Robert Eccles eds. *Networks and Organizations: Structure, Form, and Action.* Boston, Mass: Harvard Business School Press, pp. 25-56.

Gray, John V., Enno Siemsen, and Gurneeta Vasudeva. 2015. "Colocation still matters: Conformance quality and the interdependence of R&D and manufacturing in the pharmaceutical industry." *Management Science,* Vol. 61, no. 11, pp. 2760-2781.

Griffin, A., and J.R. Hauser, 1996. "Integrating R&D and marketing: A review and analysis of the literature." *Journal of Product Innovation Management: An International Publication of the Product Development & Management Association, Vol.* 13, no. 3, pp. 191-215.

Guilbeault, Douglas., Joshua Becker, and Damon Centola, 2018. "Complex Contagions: A Decade in Review." In Lehmann, S., Ahn, Y.-Y., ed., *Complex Spreading Phenomena in Social Systems.* Switzerland: Springer, pp. 3–25.

Gupta, Anil K., and Vijay Govindarajan. 2000 "Knowledge flows within multinational corporations." *Strategic Management Journal,* Vol. 21, no. 4, pp. 473-496.

Gutwin, Carl, Saul Greenberg, Roger Blum, Jeff Dyck, Kimberly Tee, and Gregor McEwan. 2008. "Supporting Informal Collaboration in Shared-Workspace Groupware." *J. UCS,* Vol. 14, no. 9, pp. 1411-1434.

Hambrick, Donald C., Theresa Seung Cho, and Ming-Jer Chen. 1996. "The influence of top management team heterogeneity on firms' competitive moves." *Administrative Science Quarterly, Vol.* 41, no. 4, pp. 659-684.

Hoegl, Martin, and Luigi Proserpio. 2004. "Team member proximity and teamwork in innovative projects." *Research Policy,* Vol. 33, no. 8, pp. 1153-1165.

Hogg, Michael A., and Deborah I. Terry. 2000. "Social identity and self-categorization processes in organizational contexts." *Academy of Management Review,* Vol. 25, no. 1, pp. 121-140.

Holland, Sarah, Kevin Gaston, and Jorge Gomes. 2000. "Critical success factors for cross-functional teamwork in new product development." *International Journal of Management Reviews, Vol.* 2, no. 3 (Fall), pp. 231-259

Jackson, Susan E. 1992. "Consequences of group composition for the interpersonal dynamics of strategic issue processing." *Advances in Strategic Management,* Vol. 8, no. 3, pp. 345-382.

Jackson, Susan E., and Aparna Joshi. 2004. "Diversity in social context: a multi-attribute, multilevel analysis of team diversity and sales performance." *Journal of Organizational Behavior: The International Journal of Industrial, Occupational and Organizational Psychology and Behavior,* Vol. 25, no. 6, pp. 675-702.

Jackson, Susan E., Karen E. May, Kristina Whitney, Richard A. Guzzo, and Eduardo Salas. 1995. "Understanding the dynamics of diversity in decision-making teams." *Team Effectiveness and Decision Making in Organizations,* Vol. 204, pp. 204-261.

Jewels, Tony, Alan Underwood, and Carmen de Pablos Heredero. 2003. "The role of informal networks in knowledge sharing." *European Conference on Information Systems Proceedings* (Naples, June 16-21, 2003).

Jones, Adrian. 2006. "Multidisciplinary team working: Collaboration and conflict." *International Journal of Mental Health Nursing,* Vol. 15, no. 1, pp. 19-28.

Kale, Prashant, Harbir Singh, and Howard Perlmutter. 2000. "Learning and protection of proprietary assets in strategic alliances: Building relational capital." *Strategic Management Journal,* Vol. 21, no. 3, pp. 217-237.

Kaufmann, Lutz, and Claudia M. Wagner. 2017. "Affective diversity and emotional intelligence in cross-functional sourcing teams." *Journal of Purchasing and Supply Management,* Vol. 23, no. 1, pp. 5-16.

Kaye, Adam, and Tessa Cook. 2015. "Improving radiologist-IT staff communications and collaboration through a shadowing project." *Journal of Digital Imaging,* Vol. 28, no. 4, pp. 433-438.

Kearney, Eric, and Diether Gebert. 2009. "Managing diversity and enhancing team outcomes: the promise of transformational leadership." *Journal of Applied Psychology,* Vol. 94, no. 1, pp. 77-89.

Kearney, Eric, Diether Gebert, and Sven C. Voelpel. 2009. "When and how diversity benefits teams: The importance of team members' need for cognition." *Academy of Management Journal,* Vol. 52, no. 3 pp. 581-598.

Kijkuit, Bob, and Jan van den Ende. 2010. "With a little help from our colleagues: A longitudinal study of social networks for innovation." *Organization Studies,* Vol. 31, no. 4, pp. 451-479.

Knouse, Stephen B., and Mickey R. Dansby. 1999. "Percentage of work-group diversity and work-group effectiveness." *The Journal of Psychology,* Vol. 133, no. 5, pp. 486-494.

Kraut, Robert E., and Lynn A. Streeter. 1995. "Coordination in software development." *Communications of the ACM,* Vol. 38, no. 3, pp. 69-82.

Kraut, Robert E., Robert S. Fish, Robert W. Root, and Barbara L. Chalfonte. 1990. "Informal communication in organizations: Form, function, and technology." *Human Reactions to Technology: Claremont Symposium on Applied Social Psychology,* pp. 145-199.

Kramer, Roderick M. 2010. "Collective trust within organizations: Conceptual foundations and empirical insights." *Corporate Reputation Review,* Vol. 13, no. 2, pp. 82-97.

Koohang, Alex, Joanna Paliszkiewicz, and Jerzy Goluchowski. 2017. "The impact of leadership on trust, knowledge management, and organizational performance: A research model." *Industrial Management & Data Systems,* Vol. 117, no. 3, pp. 521-537.

Latham, Gary P., Dawn C. Winters, and Edwin A. Locke. 1994. "Cognitive and motivational effects of participation: A mediator study." *Journal of Organizational Behavior,* Vol. 15, no. 1, pp. 49-63.

Lee, Dabae, Yeol Huh, and Charles M. Reigeluth. 2015. "Collaboration, intragroup conflict, and social skills in project-based learning." *Instructional Science,* Vol. 43, no. 5, pp. 561-590.

Lee, Sung-Mahn, and Juneseuk Shin. 2017. "A Path to Collaborative Innovation Through Internal Boundary Breaking: Open innovation tools applied within the organization helped LG Chem Research Park build a culture that nurtures collaborative innovation." *Research-Technology Management,* Vol. 60, no. 4, pp. 26-32.

Leenders, Mark AAM, and Berend Wierenga. 2002. "The effectiveness of different mechanisms for integrating marketing and R&D." *Journal of Product Innovation Management: An International Publication of the Product Development & Management Association,* Vol. 19, no. 4, pp. 305-317.

Liu, Chia-Ling, Pervez N. Ghauri, and Rudolf R. Sinkovics. 2010. "Understanding the Impact of Relational Capital and Organizational Learning on Alliance Outcomes." *Journal of World Business,* Vol. 45, no. 3, pp. 237–249.

Lovelace, Kay, Debra L. Shapiro, and Laurie R. Weingart. 2001. "Maximizing cross-functional new product teams' innovativeness and constraint adherence: A conflict communications perspective." *Academy of Management Journal,* Vol. 44, no. 4, pp. 779-793.

Maurer, Indre, Vera Bartsch, and Mark Ebers. 2011. "The value of intra-organizational social capital: How it fosters knowledge transfer, innovation performance, and growth." *Organization Studies,* Vol. 32, no. 2, pp. 157-185.

McGrath, Rita Gunther. 2001. "Exploratory learning, innovative capacity, and managerial oversight." *Academy of Management Journal,* Vol. 44, no. 1, pp. 118-131.

Mejia, David A., Jesus Favela, and Alberto L. Moran. 2008. "Augmenting informal collaboration in hospitals through pervasive computing." *2008 Second International Conference on Pervasive Computing Technologies for Healthcare* (Tampere, Finland, Jan 30 – Feb 1, 2008), pp. 237-240.

Milliken, Frances J., and Luis L. Martins. 1996. "Searching for common threads: Understanding the multiple effects of diversity in organizational groups." *Academy of Management Review,* Vol. 21, no. 2, pp. 402-433.

Mom, Tom J. M., Pepijn van Neerijnen, Patrick Reinmoeller, and Ernst Verwaal. 2015. "Relational Capital and Individual Exploration: Unravelling the Influence of Goal Alignment and Knowledge Acquisition." *Organization Studies,* Vol. 36, no. 6, pp. 809–829.

Montoya, R. Matthew, and Todd L. Pittinsky. 2011. "When increased group identification leads to outgroup liking and cooperation: The role of trust." *The Journal of Social Psychology,* Vol. 151, no. 6, pp. 784-806.

Moran, Peter. 2005. "Structural vs. relational embeddedness: Social capital and managerial performance." *Strategic Management Journal,* Vol. 26, no. 12, pp. 1129-1151.

Muller, Michael J., Werner Geyer, Beth Brownholtz, Eric Wilcox, and David R. Millen. 2004. "One-hundred days in an activity-centric collaboration environment based on shared objects." *Proceedings of the SIGCHI Conference on Human Factors in Computing Systems* (Vienna, Austria, April, 2004), New York, NY: Association for Computing Machinery, pp. 375-382.

Olson, Bradley J., Satyanarayana Parayitam, and Yongjian Bao. 2007. "Strategic decision making: The effects of cognitive diversity, conflict, and trust on decision outcomes." *Journal of Management,* Vol. 33, no. 2, pp. 196-222.

Olson, Eric M., Orville C. Walker Jr, and Robert W. Ruekert. 1995. "Organizing for effective new product development: The moderating role of product innovativeness." *Journal of Marketing, Vol.* 59, no. 1, pp. 48-62.

Patrashkova, Ralitza R. and Sara McComb. 2004. " Exploring why more communication is not better: insights from a computational model of cross-functional teams." *Journal of Engineering and Technology Management,* Vol. 21, no. 1-2. pp 83-114.

Pelled, Lisa Hope. 1996. "Relational demography and perceptions of group conflict and performance: A field investigation." *International Journal of Conflict Management,* Vol. 7, no. 3, pp. 230-246.

Pelled, Lisa Hope, Kathleen M. Eisenhardt, and Katherine R. Xin. 1999. "Exploring the black box: An analysis of work group diversity, conflict and performance." *Administrative Science Quarterly,* Vol. 44, no. 1, pp. 1-28.

Peterson, S. J., and F. Luthans. 2006. "The impact of financial and nonfinancial incentives on business-unit outcomes over time." *Journal of Applied Psychology,* Vol. 91, no. 1, pp. 156-165.

Phillips, Katherine W., and Denise Lewin Loyd. 2006. "When surface and deep-level diversity collide: The effects on dissenting group members." *Organizational Behavior and Human Decision Processes,* Vol. 99, no. 2, pp. 143-160.

Pinto, Mary Beth, Jeffrey K. Pinto, and John E. Prescott. 1993. "Antecedents and consequences of project team cross-functional cooperation." *Management Science,* Vol. 39, no. 10, pp. 1281-1297.

Proehl, Rebecca A. 1996. "Enhancing the effectiveness of cross-functional teams." *Leadership & Organization Development Journal,* Vol. 17, no. 5, pp. 3-10.

Randel, Amy, and Kimberly S. Jaussi. 2003. "Functional background identity, diversity, and individual performance in cross-functional teams." *Academy of Management Journal,* Vol. 46, no. 6, pp. 763-774.

Riege, Andreas. 2005. "Three-dozen knowledge-sharing barriers managers must consider." *Journal of Knowledge Management,* Vol. 9, no. 3, pp. 18-35.

Schütz, Peter, and Brian Bloch. 2006. "The "silo-virus": diagnosing and curing departmental groupthink." *Team Performance Management: An International Journal,* Vol. 12, no. 1, pp. 31-43.

Sethi, Rajesh, Daniel C. Smith, and C. Whan Park. 2001. "Cross-functional product development teams, creativity, and the innovativeness of new consumer products." *Journal of Marketing Research,* Vol. 38, no. 1, pp.73-85.

Sessoms, Gail. 2017. "What Are Organizational Silos?." Your Business. <https://yourbusiness.azcentral.com/organizational-silos-8237.html>. [May 04, 2020].

Shin, Shung J., and Jing Zhou. 2007. "When is educational specialization heterogeneity related to creativity in research and development teams? Transformational leadership as a moderator." *Journal of Applied Psychology,* Vol. 92, no. 6. pp. 1709-1721.

Somech, Anit. 2006. "The effects of leadership style and team process on performance and innovation in functionally heterogeneous teams." *Journal of Management,* Vol. 32, no. 1, pp. 132-157.

Song, Dongmei. 2009. "The tacit knowledge-sharing strategy analysis in the project work." *International Business Research,* Vol. 2, no. 1, pp. 83-85.

Song, X. Michael, and Mark E. Parry. 1997. "A cross-national comparative study of new product development processes: Japan and the United States." *Journal of Marketing,* Vol. 61, no. 2, pp. 1-18.

Stewart, Marcus M., and Olenda E. Johnson. 2009. "Leader—Member exchange as a moderator of the relationship between work group diversity and team performance." *Group & Organization Management,* Vol. 34, no. 5, pp. 507-535.

Stone, Florence. 2004. "Deconstructing silos and supporting collaboration." *Employment Relations Today,* Vol. 31, no. 1, pp. 11-18.

Suprapto, Mohammad, Hans LM Bakker, Herman G. Mooi, and Wim Moree. 2015. "Sorting out the essence of owner–contractor collaboration in capital project delivery." *International Journal of Project Management,* Vol. 33, no. 3, pp. 664-683.

Swamidass, Paul M., and M. Dayne Aldridge. 1996. "Ten rules for timely task completion in cross-functional teams." *Research Technology Management,* Vol. 39, no. 4, pp. 12.

Sy, Thomas, and Stéphane Côté. 2004. "Emotional intelligence: A key ability to succeed in the matrix organization." *Journal of Management Development,* Vol. 23, no. 5, pp. 437-455.

Tajfel, Henri. 1982. "Social psychology of intergroup relations." *Annual Review of Psychology,* Vol. 33, no. 1, pp. 1-39.

Tekleab, Amanuel G., Ayse Karaca, Narda R. Quigley, and Eric W.K. Tsang. 2016. "Re-examining the functional diversity–performance relationship: The roles of behavioral integration, team cohesion, and team learning." *Journal of Business Research,* Vol. 69, no. 9, pp. 3500-3507.

Temby, Owen, Jean Sandall, Ray Cooksey, and Gordon M. Hickey. 2017. "Examining the role of trust and informal communication on mutual learning in government: the case of climate change policy in New York." *Organization & Environment,* Vol. 30, no. 1, pp. 71-97.

Terman, Jessica N., Richard C. Feiock, and Jisun Youm. 2020. "When collaboration is risky business: The influence of collaboration risks on formal and informal collaboration." *The American Review of Public Administration,* Vol. 50, no. 1, pp. 33-44.

Thomson, Ann Marie, James L. Perry, and Theodore K. Miller. 2009. "Conceptualizing and measuring collaboration." *Journal of Public Administration Research and Theory,* Vol. 19, no. 1, pp. 23-56.

Tsai, Kuen-Hung, and Teresa Tiaojung Hsu. 2014. "Cross-Functional collaboration, competitive intensity, knowledge integration mechanisms, and new product performance: A mediated moderation model." *Industrial Marketing Management.* Vol. 43, no. 2, pp. 293-303.

Tuer, Frances L. 2013. "Exploring the Role of Perceptions of Trustworthiness in Heterogeneous Teams." PhD diss., McMaster University.

Turkulainen, Virpi, and Mikko Ketokivi. 2012. "Cross-functional integration and performance: what are the real benefits?" *International Journal of Operations & Production Management, Vol.* 32, no. 4, pp. 447-467.

Troy, Lisa C., Tanawat Hirunyawipada, and Audhesh K. Paswan. 2008. "Cross-functional integration and new product success: an empirical investigation of the findings." *Journal of Marketing,* Vol. 72, no. 6, pp. 132-146.

Van den Bulte, Christophe, and Rudy K. Moenaert. 1998. "The effects of R&D team co-location on communication patterns among R&D, marketing, and manufacturing." *Management Science,* Vol. 44, no. 11-part-2, pp. S1-S18.

Van Knippenberg, Daan, and Michaela C. Schippers. 2007. "Work group diversity." *Annual Review of Psychology,* Vol. 58, pp. 515-541.

Van Leeuwen, Esther, Susanne Täuber, and Kai Sassenberg. 2011. "Knocking on the outgroup's door: Seeking outgroup help under conditions of task or relational conflict." *Basic and Applied Social Psychology,* Vol. 33, no. 3, pp. 266-278.

Vatanpour, Hossein, Atoosa Khorramnia, and Naghmeh Forutan. 2013. "Silo effect a prominence factor to decrease efficiency of pharmaceutical industry." *Iranian Journal of Pharmaceutical Research,* Vol. 12, no. 12, pp. 207-216.

Waller, Mary J., George P. Huber, and William H. Glick. 1995. "Functional background as a determinant of executives' selective perception." *Academy of Management Journal,* Vol. 38, no. 4, pp. 943-974.

Webber, Sheila Simsarian. 2002. "Leadership and trust facilitating cross-functional team success." *Journal of Management Development,* Vol. 21, no. 3, pp. 201-214.

Webber, Sheila Simsarian. 2008. "Development of cognitive and affective trust in teams: A longitudinal study." *Small Group Research,* Vol. 39, no. 6, pp. 746-769.

Webber, Sheila Simsarian, and Lisa M. Donahue. 2001. "Impact of highly and less job-related diversity on work group cohesion and performance: A meta-analysis." *Journal of Management,* Vol. 27, no. 2, pp. 141-162.

Wei, Leong Teen., and Rashad Yazdanifard. 2014. "The impact of positive reinforcement on employees' performance in organizations." *American Journal of Industrial and Business Management, Vol.* 4, no. 1, pp. 9-12.

West, Michael A. 1990. "The social psychology of innovation in groups." Chichester: John Wiley & Sons, Ltd.

West, Michael A. 2002. "Sparkling fountains or stagnant ponds: An integrative model of creativity and innovation implementation in work groups." *Applied Psychology,* Vol. 51, no. 3, pp. 355-387.

West, Michael A., and Michaela Wallace. 1991. "Innovation in health care teams." *European Journal of Social Psychology,* Vol. 21, no. 4, pp. 303-315.

Whittaker, Steve, David Frohlich, and Owen Daly-Jones. 1994. "Informal Workplace Communication: What is it Like and How Might We Support It." *Proceedings of the SIGCHI Conference on Human Factors in Computing Systems* (Boston, April 24-28, 1994), pp. 131-137.

Widén-Wulff, Gunilla, and Mariam Ginman. 2004. "Explaining knowledge sharing in organizations through the dimensions of social capital." *Journal of Information Science,* Vol. 30, no. 5, pp. 448-458.

Williams, Robert J., James J. Hoffman, and Bruce T. Lamont. 1995. "The influence of top management team characteristics on M-form implementation time." *Journal of Managerial Issues,* Vol. 7, no. 4, pp. 466-480.

Wu, Daoyou, Zhongju Liao, and Juanlan Dai. 2015. "Knowledge heterogeneity and team knowledge sharing as moderated by internal social capital." *Social Behavior and Personality: An International Journal,* Vol. 43, no. 3, pp. 423-436.

Wulf, Anna, and Lynne Butel. 2017. "Knowledge sharing and collaborative relationships in business ecosystems and networks." *Industrial Management & Data Systems,* Vol. 117, no. 7, pp. 1407-1425.

Xin, Sufei, Ziqiang Xin, and Chongde Lin. 2016. "Effects of trustors' social identity complexity on interpersonal and intergroup trust." *European Journal of Social Psychology,* Vol. 46, no. 4, pp. 428-440.

Yong, Kevyn, Stephen J. Sauer, and Elizabeth A. Mannix. 2014. "Conflict and creativity in interdisciplinary teams." *Small Group Research,* Vol. 45, no. 3, pp. 266-289.

Zolin, Roxanne, Pamela J. Hinds, Renate Fruchter, and Raymond E. Levitt. 2004. "Interpersonal trust in cross-functional, geographically distributed work: A longitudinal study." *Information and Organization,* Vol. 14, no. 1, pp. 1-26.

Zornoza, Ana., Virginia Orengo, and Vicente Peñarroja. 2009. "Relational Capital in Virtual Teams: The Role Played by Trust." *Social Science Information,* Vol. 48, no. 2, pp. 257–81.

IV. LERA Best Papers XVII: Wages

Unions' Effects on Wage Inequality in Japan

KYOKO SUZUKI
University of Tokyo

Abstract

There has been no agreed conclusion on the union wage effect in Japan, with most research indicating that there is only a minimal effect. This paper re-evaluates the effect of unions on wage levels and distributions in the Japanese labor market, using surveys conducted by RENGO-RIALS, an affiliation of the largest national trade union center in Japan. The research indicates that most of the observed union wage gap comes from differences in firm size and differences in worker characteristics, especially tenure, between unionized and nonunionized firms, confirming no marginal effect of unions independent from these factors. However, unions exercise their influence by maintaining the wage structure that unions have long advocated, rather than simply increasing the wage level. In unionized firms, wages increase with age/tenure, and the slope of the wage curve becomes steeper after age 40. As for wage distribution, unions do not reduce the overall wage dispersion of male workers within unionized firms but reduce wage inequality within the same age groups. As for female workers, inequality is consistently larger in unionized firms than in nonunionized firms, and the gap has continued to expand in recent years. The results show that unions exert their influence by maintaining the wage structure that they have promoted. While they have been successful in implementing union wage policy, their traditional concept of fairness is challenged.

Introduction

The Union Effects on Wages

What effect do labor unions have on the Japanese labor market? The effects of labor unions are manifold, including wages, turnover, employment, and productivity. Researchers have tackled the question "What do unions do?" from various perspectives (Freeman and Medoff 1984; Bennett and Kaufman 2007). In the United States and other western countries, union effects on wages have been extensively discussed, and there have been generally agreed conclusions: unions increased wages, reduced the returns on worker characteristics, and compressed the wage structure, leading to less inequality in the labor market. Research indicates that union decline has been an important factor for widening wage inequality since the 1990s (DiNardo, Fortin, and Lemieux 1996, Fortin and Lemieux 1997; Card 2001).

However, research on union effects in Japan has been scarce, and there has been no clear conclusion on the effect of labor unions on wages. Several studies indicated that unions had zero or even negative impact on wages until the 1990s (Tachibanaki and Noda 2000; Todate 2010). Given that labor unions engage in the famous annual wage bargaining ("*Shunto*," the spring wage offensive) on a nationwide scale, it is puzzling why previous studies do not show the impact of unions on wages. In the 2000s, some research suggested that unions may have positively impacted wages during the long period of the recession. However, there have still been very few discussions on wage dispersion and union effects on the widening inequality in the labor market. One of the reasons for this scarcity of research is the lack of appropriate data in Japan. Although the unionization rate has now fallen to 17% of the employed population, this does not mean the decline of unions, especially in terms of the relative rank of union members in the labor market. Therefore, this research tries to re-evaluate the union effects on wages in Japan by using newly available data and applying different analytical models from those in previous studies.

Developments in the Japanese Labor Market

There are two aspects we should remember in considering union effects in the recent developments in the Japanese labor market. First, wages have been stagnant in the last 30 years. Figure 1 shows average annual wages for full-time employees, with the lines representing the changes of wages indexed to 1990. While average wages have steadily increased in other OECD countries including the United States, Japan is the only economy that has experienced consistent wage stagnation.

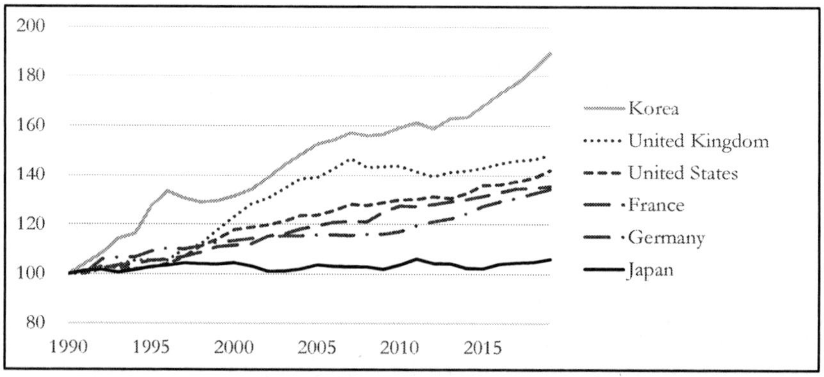

Source: OECD.stats, Average Annual Wages, 2019 constant prices.

Figure 1. Average annual wages for full-time workers (indexed to 1990).

"The second aspect is that there has been a steady rise of nonstandard types of employees. In the Japanese employment system, the terms "regular" *(seiki)* and "nonregular" *(hiseiki)* partially correspond with the terms "standard" and "nonstandard" in other employment systems but have distinctive characteristics and connotations of their own. I have therefore chosen to use them here rather than the more widely used terminology. Figure 2 shows the number of regular and nonregular employees. While regular employees have slightly decreased in the last 30 years, nonregular employees have rapidly increased and reached around 40% of the total employees.

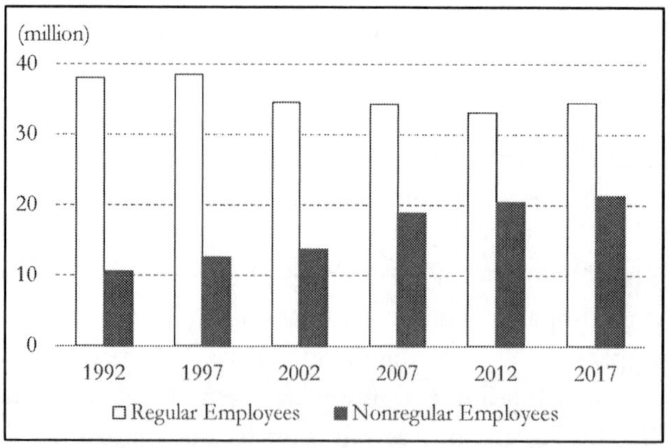

Source: e-stat, "Employment Status Survey."

Figure 2. Number of people employed in Japan (regular and nonregular employment).

There are significant disparities in labor conditions between these two types of employees, including wages, benefits, and job security. While those in regular employment are offered seniority-based wages, job security, and long-term employment, nonregular employment is confined to jobs with low wages and less security (Gordon 2017). It is more important to note that the wages of nonregular employees do not increase no matter how many years they work for the same company. While many nonregular employees perform the same tasks and work for the same hours as regular employees, their wages will not increase in the same way as regular employees even if they continue to work in the same company for many years (Osawa 2013). The distinction between these two categories lies in whether or not an employee is granted the status of a corporate community member. As already mentioned, nonregular employment has rapidly expanded since the 1990s, meaning that the share of low-wage workers in the labor market has increased. Therefore, if we look at the overall Japanese labor market, many workers have experienced even worse than their wages "being stagnant." In considering union effects, it is essential to relate the unions to these trends.

Labor Market Segments and Union Membership

The increase of nonregular employees has a significant impact not only on lowering average wages but also on union presence. The fact that labor unions in Japan have been organized on a company-by-company basis, and have only allowed regular employees to join, is believed to be the cause of this growing faultline. Figure 3 illustrates how the Japanese labor market has segmented. The labor market consists of unionized firms and nonunionized firms (solid vertical lines in the figure). The labor market is also divided by the type of employment, i.e., regular employment and nonregular employment (horizontal line in the figure). In Japan, labor unions are organized by companies, and people hired by unionized companies as regular employees automatically become union members. However, nonregular employees have long been excluded from unions even when they are hired by unionized firms. As a result, only segment (A) in Figure 3 corresponds to union members.[1]

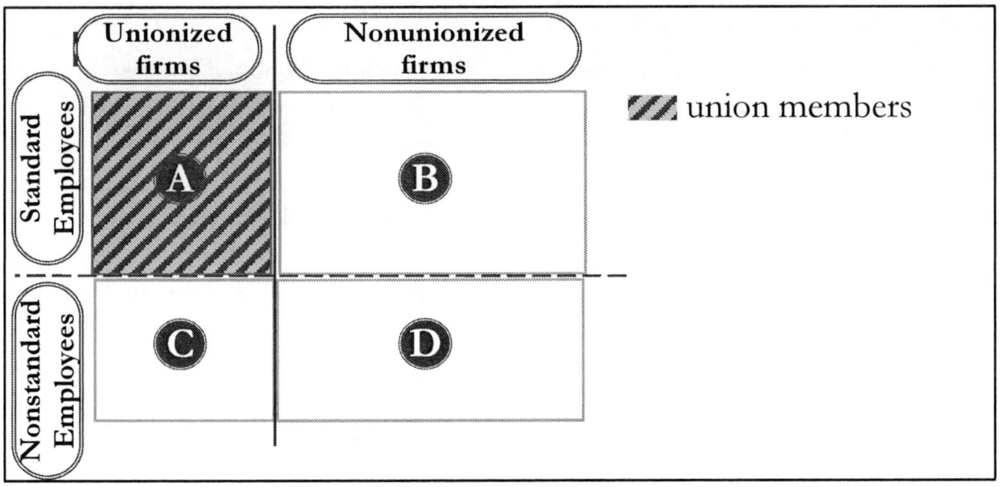

Figure 3. Labor market segments and union membership in Japan.

There are two aspects to the union effects on the Japanese labor market: one is the aspect that divides union members and nonmembers within regular employment (A vs. B). The other aspect divides regular employees (A), who are eligible for union membership, and nonregular employees, who are not eligible for it (A vs. C&D). The latter aspect is critical in evaluating union effects, as this linkage between union membership and type of employment itself suggests that labor unions play a significant role in shaping disparities between regular and nonregular employment. However, it is difficult to measure its impact, because the categories of union membership overlap with the categories of employment type, making it hard to distinguish those two effects. On the other hand, the former aspect of union effects (A vs. B) has been widely discussed in previous studies. Therefore, this article also examines the union effects within regular employment, as most previous studies did.

Contrary to common belief, the division between regular and nonregular employment is not the most prominent faultline in the Japanese labor market. A previous study estimated the latent structure of the Japanese labor market and found that it is composed of two heterogeneous segments with different wage-determining systems, just as suggested by the dual labor market theory (Suzuki 2020). However, the division between the two sectors does not entirely correspond to the division between regular and nonregular employees. The faultline extends into the regular employees: a quarter of the regular employees are indistinguishable from nonregular employees in terms of the wage-determining system. Therefore, it is important to examine what causes this segmentation among regular employees. Since the previous study (Suzuki 2020) does not investigate the effect of labor unions due to the lack of information on union membership in the data, this paper will examine the division among the regular employees using a survey containing the necessary information on unions.

Union Wage Policies in Japan

It is crucial to understand for what labor unions have campaigned in order to choose criteria for evaluating union effects. There are some differences in union wage policy between western countries and Japan. Unions in the United States and other countries have promoted policies that seek to standardize wages within and across firms and establishments. (Freeman and Medoff 1984). This policy aims to reduce management's discretion on wage determination, leading to narrower wage gaps among different demographic and skill groups. Freeman (1980) and subsequent research confirmed that unions compress wage distribution within union sectors, and this effect was significant enough to improve overall wage equality in the labor market (Freeman 1980; Card, Lemieux and Riddell 2004).

Unions' wage policy in Japan is quite different. Unions have long promoted a wage structure in which wages increase with age/tenure so that workers can support their families with the income of the male workers. Wages are determined according to workers' membership of the company, which has little relevance to each worker's occupation. Unions have advocated proper "discrimination" of wages according to workers' contribution, often leading to disparities by age and gender (Nimura 1994). Japanese workers consider it fair if employees are treated uniformly within the same age groups in the same company, no matter what kind of work they undertake. This unique sense of fairness that unions have advocated stems from the history of labor unions in Japan.

A key feature of Japanese unions is that they have been organized at the enterprise level and not at the occupation or industry level. These "enterprise unions" are also "mixed occupation unions" which include both blue-collar and white-collar workers. According to Nimura (1994), these characteristics were formed in the 1950s and have left a significant impact on unions' policies. For example, unions have campaigned for abolishing differences between blue- and white-collar workers in terms of status as well as wages and other benefits and demanded to integrate the two different compensation systems. As a result, blue-collar workers came to have a similar wage structure as white-collar workers by the end of the 1960s (Nimura 1994).

The essential value advocated through this campaign was "equality as a member of the company." However, this equality was applied only for regular employees and was also limited to male workers. When the economy requires more flexibility in workforce management, it has been female workers who bear its cost (Osawa 2001). Nonregular employment has been expanding since the 1980s, and married women have filled these positions at the minimum wage. Unions have repeatedly rejected the concept of equal pay for equal work by promoting wages that increase with age and tenure.

While the union-promoted wage structure increases the wage gap between different age groups, it reduces the dispersion of wages among workers of the same age group within the same company. This kind of wage structure may still be maintained in some unionized companies. Since the 1990s, Japan has suffered from a prolonged recession, and many companies were forced to cut wages. During this period, enterprise unions made every effort to maintain the existing wage structure within each company, while managements have tried to introduce the performance-based system to modify seniority-based wage systems. If data shows a wage distribution corresponding to union wage policy across the entire labor market, it could be said that there is a union effect on wages in Japan.[2]

Measuring Union Wage Effect

Statistical models to estimate union effects on wages have long been discussed. Given that union wage policies are different in each country, it is important to choose the appropriate model capturing union effects according to union wage policy. In Japan, there has been no agreed conclusion on whether there is any union effect on wages (Todate 2010). It was not thought that union membership would have any wage-increasing effect until the 1990s(Tachibanaki and Noda 2000; Noda 1997). Some research indicated that unions raised wages for both men and women in the early 2000s (Hara and Kawaguchi 2008; Nitta and Shinozaki 2008) and unions raised wages only for men in the late 2000s (Nitta and Shinozaki 2008; Tsuru 2010). Hara and Kawaguchi (2008) suggested that positive union effects in the 2000s may have occurred because wages in unionized firms remained relatively high during the prolonged recession period of the 2000s. However, there are differences in variables and analytical models among those studies, and they also rely on surveys with relatively small sample size, it is still unclear whether union effects could be observed when those factors are fully taken into account.[3] Also, there are only a few discussions on union effects in the 2010s, which requires further studies on union effects in Japan.

One of the reasons why previous studies could not capture union effects lies in their analytical model. Most of the previous studies have adopted an approach of estimating the union effects, using Ordinary Least Squares (OLS) as follows:

$$lnW = aU + \sum_1^k b_k X_k \tag{1}$$

where lnW is the log wage, U is a binary variable representing union membership, and X_k represent other variables for individual characteristics. This model estimates the marginal effect of union membership, controlling for other factors, such as gender, age, and company size. Since these factors are strongly related to union membership in Japan, the simple OLS model is not suitable for capturing the union effect. Union effect is no longer observed after controlling for them (Tachibanaki and Noda 2000; Nitta and Shinozaki 2008). However, the marginal effect is not the only aspect we should consider. The relationships between union status and other factors are also important to understand the role of unions in generating inequality.

In the United States and other countries, it is widely accepted that unions positively impact wages they negotiate. A widely used model is to estimate two separate wage functions for the unionized and nonunionized firms, assuming that wage-determining systems (represented as wage function) are different between the two sectors. The wage functions are written as:

$$lnW^u = \sum_k b_k^u X_k^u, \qquad \text{and} \qquad lnW^n = \sum_k b_k^n X_k^n \tag{2}$$

where index "u" and "n" represent unionized and nonunionized firm, respectively, lnW^u, lnW^n are log wage, b_k^u, b_k^n are coefficients of wage function, and X_k^u represent variables for individual characteristics. The difference between the two equations represents the wage gap due to different effects of worker characteristics in the two sectors (Lewis 1986).

$$\widehat{lnW^u} - \widehat{lnW^n} = \sum_k (\widehat{b_k^u} - \widehat{b_k^n}) X_k \tag{3}$$

This model is more suitable for measuring the union effect in Japan as it can take account of different effects of individual characteristics between unionized and nonunionized firms. While previous research in the United States and other countries has put much importance on the issues of unobserved characteristics (Farber 1983; Card 1996; Lemieux 1998), observable variables such as age and tenure may be more important in Japan, given the union wage policy. Therefore, this paper adopts Blinder-Oaxaca Decomposition to examine union effects, decomposing them into two components: the differences in the composition of worker characteristics and the differences in returns on worker characteristics.[4]

Methodology

Viewpoints

This paper adopts an approach different from previous studies on the following three points. First, as for the variable representing union status, organization-level information ("Is your company/establishment organized by a union?") is used rather than individual-level information ("Are you a member of a union?"). This is because union status is not determined by the choice of an individual. Workers hired by unionized firms as regular employees will automatically become union members. If workers are hired by nonunionized firms or as nonregular employees, they may not have an opportunity to become union members. For example, Hara and Kawaguchi (2008) suggest that the reason for union effects in Japan is that it is difficult for management to cut wages during recession periods if the company is unionized. To capture the impact of such management practice, union status should be defined as an organization-level variable rather than as an individual-level variable.[5]

Secondly, middle management will be included in the analysis. In most previous studies, managers have been excluded from the analysis because they are not eligible for union membership. However, given that we focus on the union effects at the organizational level, it is reasonable to assume union effects extend to the managerial class. This comes from the fact that many employees work for the same company for many years and promote to managers, which makes the wage structures of managers and nonmanagers continuous.

Thirdly, the focus of the analysis will not be limited to the marginal effect of unions but includes the broader aspects, including interaction with other factors. The relationships between union membership and other factors can be considered to be an important part of union influence.

Research Questions

The relationship between the presence of unions and wages will be examined from three perspectives. First, union effects on wage level and wage curve will be examined. While it is obvious that a large disparity in average wages is observed between unionized and nonunionized firms, the study aims to examine where these gaps come from. The Blinder-Oaxaca Decomposition will be used to analyze how various factors relate to wage gaps between the two sectors. This model decomposes the wage gaps into two components: the difference in the composition of worker characteristics and the difference in returns on them.[6] Also, the differences in wage structures between the two sectors will be analyzed. In particular, how the effect of age differs between the two groups will be examined, based on a finding by Noda (1997) that age has been an essential factor for union wage effects.[7]

The second question is how unions affect wage distribution. Previous studies in the United States, United Kingdom, and Canada have consistently found that the distribution of wages is smaller in the union sector (Freeman & Medoff 1984; Card, Lemieux and Riddell 2004). However, in Japan, little has been known about union effects on wage distribution (Hara and Kawaguchi 2008).[8] As discussed earlier, unions have advocated a wage structure that progressively increases with age/tenure and keeps wage differences within the same age group minimal. However, with the recent efforts to introduce a performance-based pay system, these characteristics may be changing. Therefore, this study will examine whether the traditional union wage structure is still maintained in unionized firms by decomposing the distribution of log wages into within- and between-age groups.

The third question is how the presence of unions affects the overall wage inequality among all regular employees. Unions increase wage inequality in the labor market by creating wage gaps between unionized and nonunionized sectors. On the other hand, unions also reduce wage inequality by promoting standardized wages within the unionized sector (Freeman 1980). Whether the effect of increasing or decreasing inequality is more significant depends on the magnitude of each effect and unionization rate. Previous research found that the equalizing effect was more significant than the de-equalizing effect in the United States from the 1970s to the 1990s, suggesting that unions made the labor market more equal (Freeman 1980; DiNardo, Fortin and Lemieux 1996; Card, Lemieux and Riddell 2004). Using the same method as Freeman (1980) and Card, Lemieux, and Riddell (2004), the paper will examine the overall union effects on wage inequality for regular employees in Japan.

Data

The data used in this analysis is the "The Survey on Work and Life of People" conducted by the RENGO Research Institute for Advancement of Living Standards (RENGO-RIALS), an affiliation of the RENGO, or Japanese Trade Union Confederation, the largest national trade union center in Japan. The survey is conducted every six months, targeting people employed in the private sector (aged 20-64) who live in the Tokyo metropolitan and the Kansai metropolitan area. The sample size is from 2,000 to 4,000 per survey. The surveys adopt a quota sampling method, and the quotas are defined by sex and age groups based on the "Employment Status Survey" by the Ministry of Internal Affairs and Communications.[9] The analyses use repeated, cross-sectional data from the 9th survey (April 2005) to the 34th survey (October 2017).[10]

This survey includes essential questions for the analyses in this paper. The survey asks two different questions on union status together with standard questions on wages and individual characteristics: one is "Are you a member of a labor union?" and the other is "Does the company you work for have a labor union?" These questions allow us to analyze the union effects with a sample of sufficient size.[11]

In the analysis, the target is restricted to regular employees in the private sector between the ages of 20 and 59. Top management (*Yakuin*) are excluded from the analysis, while middle managers are included. The dependent variable is the log of hourly wages.[12] As for union status, the question "Does the company you work for has a union?" is used. Other variables include education, company size, tenure, gender, marital status, industry, and occupation. Data with missing values are excluded from the analysis. The final sample size is 21,439, of which the descriptive statistics is provided in Appendix 1. The data is pooled for three years to ensure a sufficient size for each group: for example, data from 2005, 2006, and 2007 are merged into one group and labeled as 2005, and data from 2008, 2009, and 2010 are merged into one group and labeled as 2008. Note that there was a change in the survey method (from mail to web-based survey) in 2011.

Figure 4 (next page) shows the responses to the two questions "Does the company you work for have a union?" and "Are you a member of a labor union?" The solid line represents the percentage of people who answered, "There is a union in my company." The broken line represents the percentage of people who answered, "I am a member of the union in my company".[13] The dotted line represents the percentage of people who answered, "I am a member of a union outside the company."

The percentage of people working for unionized firms (solid line) has been hovering around 40% for both regular and nonregular employment, with no significant difference in recent years. While most of the regular employees in unionized firms are also union members, less than half of the nonregular employees in unionized firms are union members. Some companies started to organize nonregular employees, but most unions are slow and reluctant to include them. The percentage of people who join external unions is tiny for both regular and nonregular employees.

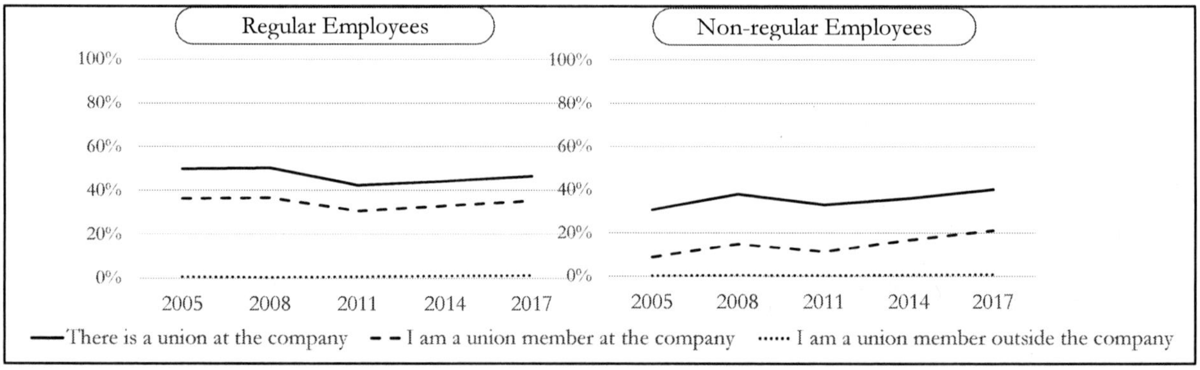

Figure 4. Union presence and union membership.

Analytical Models

Impact on Wage Level

I use the Blinder-Oaxaca Decomposition of the wage function to examine the union effects on the wage level. The model used in this analysis is called the Three-fold Decomposition (Jann 2008), which decomposes the average wage difference between the two groups into i) differences in mean covariate values (Endowment), ii) differences in the regression coefficients (Coefficients), and iii) interaction between the two (Interaction). Now, the wage functions of the unionized firms (u) and the nonunionized firms (n) are defined as follows:

$$lnW^u = \sum_k b_k^u X_k^u, \qquad lnW^n = \sum_k b_k^n X_k^n \tag{4}$$

where lnW^u, lnW^n are logs of hourly wage, b_k^u, b_k^n are coefficients of wage functions, and X_k^u represents independent variables for individual characteristics. Index "u" and "n" represent unionized firms and nonunionized firms, respectively, and k is the number of independent variables. In this case, the mean difference in outcomes of the two wage functions can be decomposed as follows:

$$E(\widehat{lnW^u}) - E(\widehat{lnW^n}) = \sum_k \widehat{b_k^u}\, E(X_k^u) - \sum_k \widehat{b_k^n}\, E(X_k^n)$$

$$= \sum_k \{\widehat{b_k^n}[E(X_k^u) - E(X_k^n)] + (\widehat{b_k^u} - \widehat{b_k^n})E(X_k^n) + (\widehat{b_k^u} - \widehat{b_k^n})[E(X_k^u) - E(X_k^n)]\} \tag{5}$$

where the three terms of equation (5) represent the following components, respectively:

Endowments: $E = \sum_k \widehat{b_k^n}[E(X_k^u) - E(X_k^n)]$

Coefficients: $C = \sum_k (\widehat{b_k^u} - \widehat{b_k^n})E(X_k^n)$

Interaction: $I = \sum_k (\widehat{b_k^u} - \widehat{b_k^n})[E(X_k^u) - E(X_k^n)]$ (6)

Effect on the Distribution of Wages

This study uses the variance of the logarithms to measure the distribution of wages in the two groups of firms. When the wage of an individual is defined as x_i, the variance of logarithm (V) is defined as follows:[14]

$$V = \frac{1}{n}\sum_{i=1}^{n}(z_i - \bar{z})^2, \quad z_i = logx_i \tag{7}$$

Now, assuming several groups, the overall variance can be decomposed into within-group variances and between-group variances. Between-group variances refer to variances caused by the average wage gap among subgroups.

Total variances : $V(s, V, \overline{W}) = \{\sum_{j=1}^{J} s_j \cdot V_j\} + \left\{\sum_{j=1}^{J} s_j \cdot \overline{W_j}^2 - (\sum_{j=1}^{J} s_j \cdot \overline{W_j})^2\right\}$ (8)

Within-group variances: $\sum_{j=1}^{J} s_j \cdot V_j$

Between-group variances: $\sum_{j=1}^{J} s_j \cdot \overline{W_j}^2 - (\sum_{j=1}^{J} s_j \cdot \overline{W_j})^2$

where *j* is an index representing subgroup (*j* =1…*J*), s_j is population share of group *j* ($\sum_{j=1}^{J} s_j = 1$), V^j is variances within group j, and $\overline{W_j}$ is an average wage for group j.

Overall Impact on Regular Employment

To estimate the overall impact on regular employment, I rely on the method used by Freeman (1980) as well as Card, Lemieux, and Riddle (2004). This method assumes a hypothetical situation in which there is no labor union, with the variance of V^n. Then the labor market will be divided into two groups of firms; one is for unionized firms, and the other is for nonunionized firms. When the wage level and dispersion in the unionized sector change, it also changes the overall dispersion of the entire labor market. If the average wage and the variance for each group are (W^u, W^n) and (V^u, V^n), respectively, the overall variance (V) is defined as follows.[15] α represents the unionized sector's share.

$$V = \alpha V^u + (1 - \alpha)V^n + \alpha(1 - \alpha)(\overline{W^u} - \overline{W^n})^2 \tag{9}$$

Equation (6) can be transformed to obtain the difference between the variance in the initial state (V^n) and variances after the change (V):

$$V - V^n = \alpha(V^u - V^n) + \alpha(1 - \alpha)(\overline{W^u} - \overline{W^n})^2 \tag{10}$$

The first term represents the change due to the difference of variance between unionized and nonunionized firms, which is negative if the unionized firms have a smaller variance and positive if they have a larger variance. The second term represents the change caused by the difference in average wages in the two groups. This term is positive as long as there is a difference in the average wage between the two groups. The change in overall variance is the sum of these two effects, whose sign depends on the size of each effect. Using the observed data for $\overline{W^n}$, $\overline{W^u}$, V^n, and V^u, the change in variance (V-V^n) due to the presence of unionized firms can be calculated. In this model, variance of nonunionized firms in the initial state and after unionization are assumed to be equal. However, in reality, there might be various heterogeneities between the two situations. An analytical model suggested by Card, Lemieux and Riddell (2004) will be used to address this issue. It first divides the sample into age and gender group, calculate union effects in each group, and then aggregate these union effects in each age and gender group.

Results

Union Effects on Wage Levels and Structures

As mentioned earlier, most previous studies did not conclude that there are union effects independent from other factors. In order to understand the reason behind this result, I estimate the basic OLS regression of wage function (1) which is commonly used in previous studies. Figure 5 summarizes the results, with the detailed estimation results shown in Appendix 2.

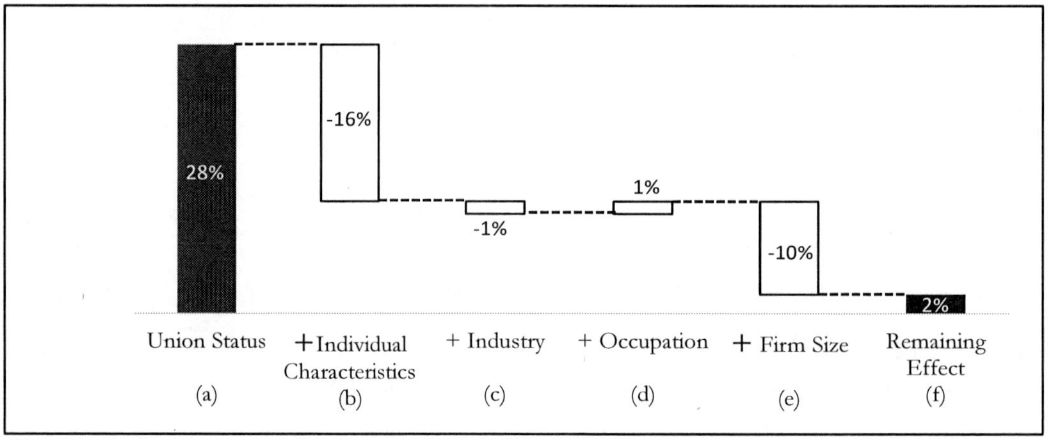

Notes:

1. Created based on the estimation results of Appendix 2.
2. Values are based on coefficients of union status and their changes among the models.
3. Values are not exponentially transformed but are interpreted as a percentage change for simplicity.

Figure 5. Effects of unions and other control variables.

Five models were estimated, with variables being added sequentially from Model (a) to Model (e). The figure shows the changes in the coefficient of the binary variable for union status. Since the dependent variable of the equation is the log of wage, the values can be interpreted as a percentage point change in the wage gaps.[16] Model (a) is the simplest regression, where log wages are regressed only on the union status. The result indicates that the average wage in unionized firms is about 28% higher than nonunionized firms. In Model (b), variables for worker characteristics such as education, work tenure, gender, and marital status are controlled. The result shows that 16 percentage points out of 28 percentage points of union effects are absorbed. Industry and occupation are further controlled in Model (c) and Model (d), respectively, but results suggest that these factors hardly influence the size of union effects. However, when the firm size is controlled in Model (e), it absorbed a further 10 percentage points of the union effects. The remaining union effect is only 2% (f) and is no longer statistically significant. The results indicate that the observed union wage gap (28%) is mostly due to differences in worker characteristics and firm size. It suggests that union membership largely overlaps with worker characteristics and firm size, and this is why previous studies could not capture union effects independent from other factors. As shown in Appendix 2, this feature did not change during the 2010s.

Next, the Blinder-Oaxaca Decomposition is applied to analyze whether the union wage gap is explained by differences in worker characteristics (Endowments) or by differences in returns to those characteristics (Coefficients).[17] Figure 6 summarizes the results, with the detailed estimation results shown in Appendix 3.

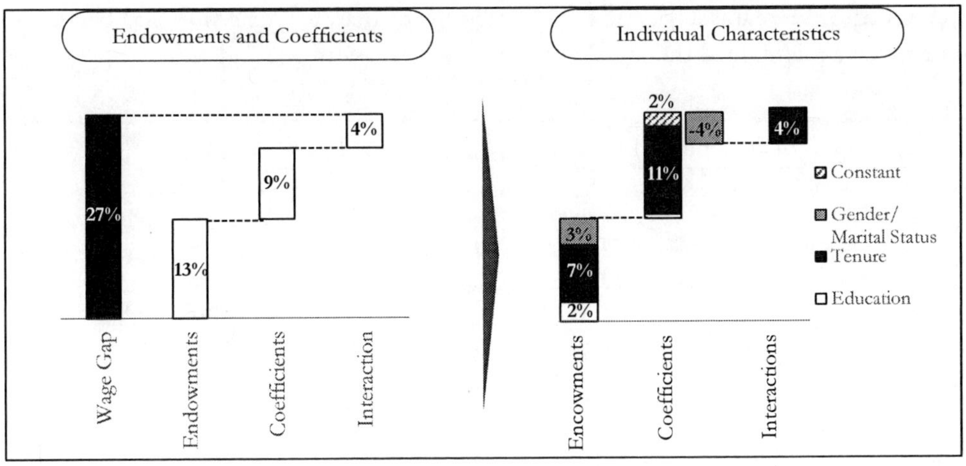

Notes:

1. Created based on the estimation results of Appendix 3.
2. Values are not exponentially transformed but are interpreted as a percentage change for simplicity.

Figure 6. Decomposition of union effects.

The bar graph on the left of Figure 6 shows an overall union wage gap of 27%. Almost half (13%) of this overall union gap comes from the difference in endowments of individual characteristics, while a minor part (9%) comes from the difference in coefficients. This suggests that unionized firms have a higher concentration of workers with characteristics leading to higher wages and have higher returns on those characteristics.

The graph on the right of Figure 6 shows the decomposition of these parts into each variable. As for the "Endowments" gap (13%), the largest part comes from tenure (years of service in the company). As for "Coefficients" (9%), most of the gap also comes from tenure. The results indicate that union wage gap arises mostly from tenure: the average tenure in unionized firms is longer, and returns on tenure are also higher than those of nonunionized firms.

Although "tenure" and "age" are two different concepts, they can be treated as almost the same thing, especially for unionized firms, where most employees work for the same company for quite a long time. Figure 7 (next page) shows the marginal effects of age and union status on wage for male and female workers based on OLS regression results. The horizontal axis represents age, and the vertical axis represents wage level.

The estimation results are shown in Appendix 4. The scale on the vertical axis is a percentage increase of wage based on the 20-24 age group in nonunionized firms.[18] Thus, the solid line represents the wage curve in the nonunionized firms, the short dotted line is the wage curve in the unionized firms, and the broken line is the wage curve in the unionized firms after controlling for firm size.[19]

As for male workers, the wage increases steadily with age, even in the nonunionized firms (solid line), reaching an almost 100% increase at age 55, compared to the wage level at age 20. Wages are consistently higher in unionized firms, and this union effect becomes larger after age 40. When controlling for firm size (broken line), there is no union effect before age 40 but union effect remains after that age. As for female workers, the trend is somewhat similar to male workers. However, the wage level in nonunionized firms does not increase with age as much as for the male workers, and the wage curve reaches a plateau at the age of 30 with just a 40% increase compared to the wage level at age 20. Wages are consistently higher for unionized firms, and the union effect becomes much larger after age 40. When controlling for firm size (broken line), the union effect after age 40 remains large and is more significant than for male workers. These results suggest that unionized firms have maintained the wage curve that increases with age and tenure.

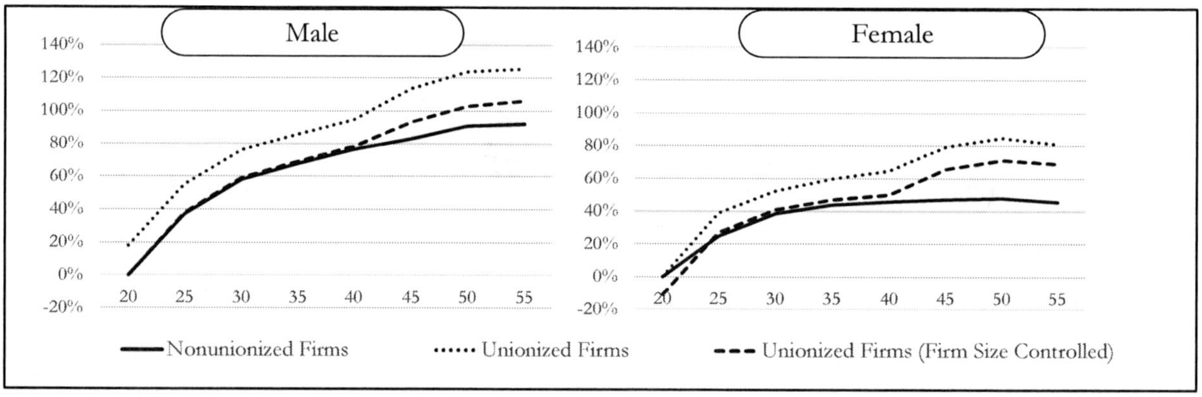

Notes:

1. Created based on the estimation results of Appendix 4.
2. The group of age 20-24 is taken as a base category.
3. "Nonunionized Firms" represents marginal effects of each age category.
4. "Unionized Firms" represents the sum of marginal effects of each age category, union effect, and interaction effects of union and each age category.
5. "Unionized Firms (Firm Size Controlled) represents the sum of marginal effects of each age category, union effect, and interaction effects of union and each age category, controlling for firm size.

Figure 7. Effects of age, unions, and firm size.

Impact on Wage Distribution

This section analyzes how unions affect wage distribution by decomposing log variance of wages into within- and between-age groups. The results by union status and gender appear in Figure 8.

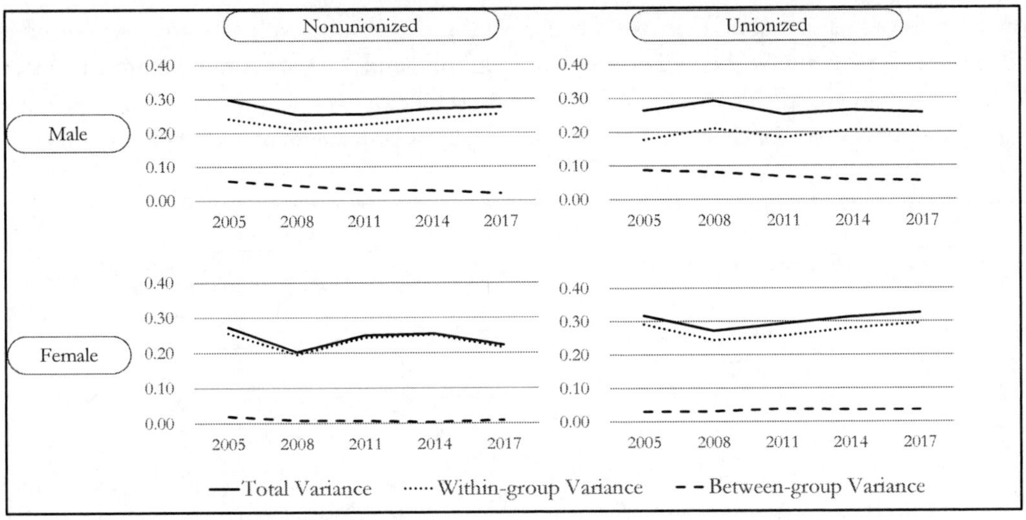

Note: Created based on Appendix 5

Figure 8. Changes in within- and between-group variances.

The solid line represents the total variance, the dotted line represents the within-group variance, and the broken line represents the between-group variance. As for male workers, while there is no significant difference in the total variance between unionized and nonunionized firms, a comparison between the two groups of firms indicates that the share of between-group variance (broken line) is larger in unionized firms than in unionized firms. This suggests that unionized firms have a wage-age curve whose slope is steeper and the dispersion within the same age group is smaller than those in nonunionized firms, which corresponds to the traditional union wage policy in Japan. However, looking at the changes over time, the between-group variance (broken line) has decreased in both groups, suggesting that the wage curve is flattening out. At the same time, while the within-group variance has increased in the nonunionized firms, it has remained almost constant in the unionized firms, suggesting that unions have managed to hold back the widening wage gap within the same age groups.

As for female workers, the total variance is consistently larger in unionized firms than nonunionized firms. However, the two groups of firms have in common that the between-group variance is minimal, and most of the total variance is accounted for by the within-group variance. This means that the wage-age curve for female workers is flat, with larger wage gaps within the same wage group than male workers. Looking at the changes over time, within-group variance in the unionized firms has consistently increased in recent years.

As we have examined, there are apparent differences in the wage distributions between the unionized and nonunionized firms. Research in the United States and other countries has consistently found that wage inequality is smaller in unionized sectors. However, this is not the case for both male and female workers in Japan. As for male workers, there is almost no difference in the total variance between the unionized and nonunionized firms. As for female workers, the total variance of wages is consistently larger in unionized firms, and the within-group variance in the same age group has continues to rise.

Overall Impact on Regular Employment

How do the unionized firms affect the overall wage distribution of regular employees? The results of the analysis appear in Figure 9. The solid line represents the total effect, i.e., the extent unionized firms increase or decrease the overall wage dispersion. This total effect is decomposed into two parts: the part coming from the difference in wage levels between the two groups (broken line), and the part coming from the difference in dispersion between the two groups (dotted line). The effects are calculated by age groups first and aggregated to adjust for the different individual characteristics of each group (Card, Lemieux and Riddell 2004).[20]

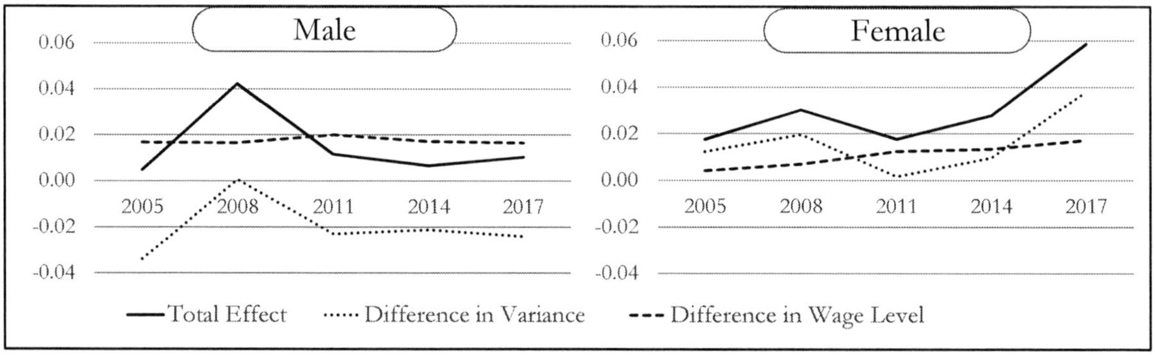

Note: Created based on Appendix 6.

Figure 9. Overall effects of unions on wage distribution of regular employees.

As for male workers, the total effect (solid line) increased in 2008 but remained close to zero after that. The variance is slightly smaller in the unionized firms than nonunionized firms, which makes the effect coming from the difference in variance of the two groups (dotted line) negative. However, this is offset by the positive effect coming from the difference in wage level between the two groups (broken line). As for female workers, the result shows that the total effect (solid line) is positive and increasing rapidly. There is also an increase in the two components of this total effect: the effect coming from the difference in variance (dotted line), and the effect coming from the average wage difference between the two groups (broken line). These two effects have combined to increase the overall wage dispersion among regular employees (solid line).

In sum, among male workers in regular employment, the unionized firms have relatively high wages and small dispersion but have little effect on the overall wage dispersion. Among female workers, however, unionized firms have relatively high wage levels and large variances among women, which also rapidly increases the overall wage dispersion.

Discussion and Conclusion

This paper analyzed the union effects on wage level and distributions of regular employees. Although there is a large union wage gap of about 30%, the largest part comes from differences in worker characteristics and firm size. Unionized firms tend to hire people with characteristics that lead to higher wages, such as male and college-graduate workers. Also, firm size is a particularly important factor in considering union effects in Japan, where unionization is concentrated in the large firm sector. There is no marginal effect of unions remaining after controlling for these factors, which confirms most previous studies on union effects in Japan (Tachibanaki and Noda 2000).

However, this does not mean that unions do not have a substantial impact on wages. The results suggest that unions exercise their influence by maintaining the wage structure that has been advocated by union policy, rather than simply increasing the wage level. Unions have long promoted a wage structure in which wages continue to increase with age/tenure over a period of more than 20 years after a worker has been hired. In unionized firms, tenure (years of service in the company) has higher returns to wage than in nonunionized firms, and the average length of tenure is also longer, and these two factors together increase workers' wages. While both unionized and nonunionized firms still maintain wage curves increasing with age, the slope becomes steeper in unionized firms after age 40. This impact can be confirmed for women in unionized firms, though the wage curve for women is flat in nonunionized firms. I suggest that maintaining this traditional wage structure of the "long-time commitment model" is the major effect of union presence.

The results also found that union effects on the distribution of wages in Japan are different from those of western countries. While studies on the United States, United Kingdom, and Canada found that unions reduce wage distribution, no such effects are observed in Japan. As for male workers, total variances are almost the same between unionized and nonunionized firms. However, the composition of variance is different between the two sectors. When decomposing variances into between- and within-age groups, results indicate that unionized firms maintain the wage structure with smaller variances within the same age groups and retain larger wage gaps between different age groups.

As for female workers, variances are consistently larger in unionized firms than in nonunionized firms, and the gap has continued to expand in recent years. This trend is related to the expanding wage gap between unionized and nonunionized firms and the rapidly increasing variance within unionized firms. Female workers used to be uniformly confined to low wages in the past. While wages for some female workers have improved in unionized sectors, most of them have still been left behind.

The results suggest that absence of marginal union effect on simple wage increase does not necessarily mean absence of union effects. The wage structure advocated by unions has been maintained in unionized firm, and this should be considered as an important union effect. Unions have been successful in defending the union wage policy and concept of fairness. However, wages and other benefits have deteriorated outside the union sector, such as people who work for small- and medium-sized firms, nonregular employees, and the self-employed. This means unions' concept of fairness have been challenged. The union effects in the entire labor market need to be investigated in future research.

Acknowledgments

The data for this secondary analysis, "The Survey on Work and Life of People" by Research Institute for Advancement of Living Standards (RENGO-REALS), was provided by the Social Science Japan Data Archive, Center for Social Research and Data Archives, Institute of Social Science, The University of Tokyo.

Endnotes

[1]Unionization rate in nonregular employment has been rapidly increasing in recent years.

[2]Industrial relations in Japan are characterized by "enterprise-unionism," consisting of labor unions organized on a company-by-company basis. Given that wages are primarily determined in a company's own wage system, wage structure can be completely different from company to company. However, unionized firms have shared wage policies and practices through collaborations such as industry-wide negotiations, which leads to similar characteristics of wage structure among them.

[3]Nitta and Shinozaki (2008) and Tsuru, Yoshinaka, and Enoki (2009) adopt annual income as the dependent variable rather than hourly wages.

[4]There are a few previous studies that adopt this approach in Japan. For example, Noda (1997) emphasizes that returns to individual characteristics differ between unionized and nonunionized firms, and Hara and Kawaguchi (2008) also conduct a decomposition analysis of the wage difference.

[5]These two answers coincide in the union shop system that is primarily adopted in large companies. In these cases, it is appropriate to treat union status as an organization-level variable. Tachibanaki and Noda (2000) discussed this point and adopted organization-level union presence as a union status variable.

[6]Hara and Kawaguchi (2008) adopt the Cotton-Neumark Decomposition to decompose the union wage gap into the difference in endowments and difference in coefficients. The effects of each separate variable are not examined.

[7]Noda (1997) confirms that the effect of tenure and age is more significant in unionized firms.

[8]Hara and Kawaguchi (2008) and Tsuru (2010) apply the DFL Decomposition method and report that labor unions have the effect of compressing the wage distribution.

[9]It should be noted that surveys targeted areas limited to Tokyo and Kansai Metropolitan Area, and were also conducted via web-based questionnaires, which may lead to potential bias in characteristics of respondents and their employers compared to samples of official statistics.

[10]Although the survey started in 2001, it started to include working hours as a survey item in 2005 and hourly wages can only be calculated after that. Also, the surveys were conducted by mail from the 1st to 20th (2010) surveys, and the 21st (April 2011) and subsequent surveys were conducted via web-based questionnaire.

[11]In contrast to the U.S., where microdata of official surveys has been available since the 1970s, large-scale surveys including union status were not available in Japan until the Japanese General Social Survey (JGSS) started in 2000. Even in research to study the union wage gap using JGSS, the sample size is between 500 and 2,500 (Nitta and Shinozaki 2008; Kawaguchi and Hara 2008).

[12]The hourly wage is calculated using annual income and weekly working hours. Since annual income and weekly work hours are asked in ranges, the central values of each range are used for calculation. Annual work hours are calculated by multiplying weekly work hours by 52. The hourly wage is then calculated by dividing the annual income by annual work hours. The survey does not include a question to determine whether the work schedule is constant throughout the year, and the calculation of hourly wage may contain some errors.

[13]Among full-time employees, most of those who work for a unionized firm but are not union members are managers who are not eligible to join a union. When managers are excluded from the sample, the rate of union membership becomes higher.

[14]Variance of log is constructed by taking the log of wages and calculating their variance. It is scale-invariant and can be compared among samples with a different unit of measurement (Allison, 1978). This index is used in previous studies that estimated the union effects in the U.S. and other countries.

[15]The basic idea of this equation is the same as that of the decomposition equation (5) in the previous section, and it corresponds to the case with only two subgroups.

[16]Values could be exponentially transformed, but instead, they are interpreted as a percentage change for simplicity.

[17]Firm size is excluded from this model, as it overlaps with the distinction between the two wage functions. Therefore, union effects include the effect of firm size in this model.

[18]Since the graph shows a wage distribution in cross-sectional data, it is not exactly a wage-age curve that workers follow over time.

[19]As shown in Appendix 4, the coefficients of age differ slightly when firm size is controlled for in nonunionized firms, but only the result of the model without firm size is shown in the graph for simplicity.

[20]Age groups in 10-year increments are used to ensure a sufficient sample size in each group.

Appendix 1. Descriptive Statistics

		Nonunionized Firms	Unionized Firms	Total
Total		11,693	9,750	21,443
		55%	45%	-
Gender	Female	30%	21%	26%
	Male	70%	79%	74%
Age	20-24	3%	5%	4%
	25-29	13%	16%	15%
	30-34	13%	12%	13%
	35-39	20%	17%	18%
	40-44	14%	13%	14%
	45-49	14%	14%	14%
	50-54	13%	14%	13%
	55-59	9%	10%	9%
Marital Status	Married	55%	64%	59%
	Unmarried	45%	36%	41%
Education	College or above	56%	69%	62%
	Middle/High/Vocational Sch	44%	31%	38%
Firm Size	- 29	48%	6%	29%
	30 - 99	19%	9%	14%
	100 - 299	10%	8%	9%
	300 - 499	7%	9%	8%
	500 - 999	7%	16%	11%
	1,000 - 2,999	7%	39%	22%
	3,000-	1%	14%	7%
Tenure	Average (year)	9.72	14.52	11.91
	s.d.	8.25	10.51	9.65
Hourly Wage (Log)	Average (yen)	7.50	7.77	7.62
	s.d.	0.52	0.55	0.55
Survey Year	2005 (2005/2006/2007)	1,303	1,295	2,598
	2008 (2008/2009/2010)	1,665	1,685	3,350
	2011 (2011/2012/2013)	3,926	2,882	6,808
	2014 (2014/2015/2016)	3,643	2,885	6,528
	2017 (2017)	1,156	1,003	2,159

Appendix 2. The Effects of Unions on Log Wage

	(1)	(2)	(3)	(4)	(5)
	Union Effects	(1) + Individual Characteristics	(2) + Industry	(3) + Occupation	(4) + Firm Size
Union Status	0.278***	0.116***	0.103***	0.116***	0.020
	(0.021)	(0.018)	(0.018)	(0.018)	(0.018)
Education (1=college or above)		0.173***	0.156***	0.110***	0.085***
		(0.007)	(0.007)	(0.007)	(0.007)
Tenure		0.030***	0.030***	0.029***	0.027***
		(0.001)	(0.001)	(0.001)	(0.001)
Tenure (squared)		-0.000***	-0.000***	-0.000***	-0.000***
		(0.000)	(0.000)	(0.000)	(0.000)
Gender (1=Female)		-0.074***	-0.076***	-0.083***	-0.081***
		(0.010)	(0.010)	(0.010)	(0.010)
Marital Status (1=Married)		0.232***	0.234***	0.197***	0.196***
		(0.008)	(0.008)	(0.008)	(0.008)
Gender * Marital Status		-0.199***	-0.200***	-0.162***	-0.159***
		(0.015)	(0.015)	(0.014)	(0.014)
Industry			Yes	Yes	Yes
			-	-	-
Occupation				Yes	Yes
				-	-
Firm Size□ Base□ below 29□					
30 - 99					0.084***
					(0.010)
100 - 299					0.113***
					(0.012)
300 - 499					0.133***
					(0.012)
500 - 999					0.161***
					(0.012)
1,000 - 2,999					0.224***
					(0.011)
3,000 or above					0.275***
					(0.017)
Survey Year□ Base: 2005□					
2008	0.029	0.034**	0.034**	0.036**	0.038**
	(0.020)	(0.017)	(0.017)	(0.016)	(0.016)
2011	0.016	0.046***	0.051***	0.056***	0.094***
	(0.017)	(0.015)	(0.015)	(0.014)	(0.014)
2014	0.043**	0.060***	0.066***	0.079***	0.116***
	(0.017)	(0.015)	(0.015)	(0.014)	(0.014)
2017	0.088***	0.100***	0.106***	0.124***	0.156***
	(0.022)	(0.019)	(0.018)	(0.018)	(0.018)
2008 * Union Status	-0.014	-0.004	-0.005	-0.004	-0.004
	(0.028)	(0.024)	(0.024)	(0.023)	(0.023)
2011 * Union Status	-0.004	0.001	0.004	0.003	0.009
	(0.025)	(0.021)	(0.021)	(0.021)	(0.021)
2011 * Union Status	-0.024	-0.010	-0.006	-0.012	-0.005
	(0.025)	(0.021)	(0.021)	(0.021)	(0.021)
2011 * Union Status	-0.027	0.006	0.010	0.005	0.017
	(0.031)	(0.027)	(0.027)	(0.026)	(0.026)
Constant	7.472***	7.032***	6.965***	7.001***	6.955***
	(0.015)	(0.015)	(0.019)	(0.020)	(0.020)
Observations	21,443	21,439	21,439	21,439	21,439
R-squared	0.059	0.312	0.322	0.356	0.372

Standard errors in parentheses
*** p<0.01, ** p<0.05, * p<0.1

Appendix 3. The Blinder-Oaxaca Decomposition

	(1) Unionized Firms	(2) Nonunionized Firms	Blinder-Oaxaca Decomposition (3) Total	(4) Endowments	(5) Coefficients	(6) Interaction
Education (1=College or above)	0.174***	0.183***		0.024***	0.005	0.001
	(0.009)	(0.010)		(0.002)	(0.007)	(0.002)
Tenure	0.027***	0.040***		0.129***	0.130***	0.064***
	(0.002)	(0.002)		(0.009)	(0.022)	(0.011)
Tenure (squared)	-0.000***	-0.000***		-0.057***	-0.021*	-0.020*
	(0.000)	(0.000)		(0.009)	(0.011)	(0.011)
Gender (1=Female)	-0.042***	-0.123***		0.004***	-0.025***	0.008***
	(0.013)	(0.015)		(0.001)	(0.006)	(0.002)
Marital Status (1=Married)	0.245***	0.192***		0.021***	-0.029***	-0.005***
	(0.011)	(0.012)		(0.002)	(0.009)	(0.001)
Gender * Marital Status	-0.242***	-0.118***		0.008***	0.014***	-0.004***
	(0.020)	(0.023)		(0.001)	(0.003)	(0.001)
Unionized Firms (Average)			7.768***			
			(0.006)			
Nonunionized Firms (Average)			7.503***			
			(0.005)			
Difference			0.265***			
			(0.007)			
Endowments			0.129***			
			(0.005)			
Coefficients			0.092***			
			(0.007)			
Interaction			0.044***			
			(0.004)			
Constant	7.109***	7.127***			0.018	
	(0.013)	(0.015)			(0.020)	
Observations	11,691	9,748	21,439	21,439	21,439	21,439
R-squared	0.196	0.359				

Standard errors in parentheses
*** $p<0.01$, ** $p<0.05$, * $p<0.1$

Appendix 4. Effects of Age, Union Status, and Firm Size

	Male		Female	
		Firm Size Controlled		Firm Size Controlled
Union Status	0.180***	0.049	-0.020	-0.112**
	(0.043)	(0.042)	(0.050)	(0.050)
Education (1=College or above)	0.177***	0.139***	0.169***	0.137***
	(0.008)	(0.008)	(0.014)	(0.014)
Age (Base: 20-24)				
25-29	0.372***	0.380***	0.248***	0.253***
	(0.035)	(0.034)	(0.044)	(0.043)
30-34	0.581***	0.591***	0.385***	0.392***
	(0.035)	(0.034)	(0.045)	(0.044)
35-39	0.679***	0.692***	0.437***	0.448***
	(0.033)	(0.033)	(0.045)	(0.044)
40-44	0.765***	0.782***	0.458***	0.483***
	(0.034)	(0.033)	(0.046)	(0.045)
45-49	0.831***	0.840***	0.472***	0.503***
	(0.034)	(0.033)	(0.046)	(0.045)
50-54	0.908***	0.912***	0.480***	0.514***
	(0.035)	(0.034)	(0.046)	(0.045)
55-59	0.920***	0.933***	0.456***	0.489***
	(0.036)	(0.035)	(0.048)	(0.047)
25-29 * Union Status	-0.028	-0.041	0.142**	0.129**
	(0.047)	(0.046)	(0.058)	(0.057)
30-34* Union Status	-0.029	-0.045	0.138**	0.129**
	(0.047)	(0.046)	(0.061)	(0.060)
35-39* Union Status	0.033	0.016	0.160***	0.133**
	(0.046)	(0.044)	(0.061)	(0.060)
40-44* Union Status	0.085*	0.049	0.189***	0.128**
	(0.047)	(0.045)	(0.064)	(0.063)
45-49* Union Status	0.125***	0.093**	0.323***	0.266***
	(0.046)	(0.045)	(0.066)	(0.065)
50-54* Union Status	0.149***	0.117**	0.369***	0.309***
	(0.047)	(0.045)	(0.067)	(0.066)
55-59* Union Status	0.154***	0.129***	0.354***	0.312***
	(0.048)	(0.047)	(0.075)	(0.074)
Firm Size (Base : below 29)				
30 - 99		0.123***		0.114***
		(0.012)		(0.020)
100 - 299		0.181***		0.131***
		(0.014)		(0.026)
300 - 499		0.206***		0.206***
		(0.014)		(0.027)
500 - 999		0.253***		0.245***
		(0.013)		(0.026)
1,000 - 2,999		0.343***		0.258***
		(0.012)		(0.022)
3,000 or above		0.411***		0.244***
		(0.018)		(0.037)
Survey Year (Base: 2005)				
2008	0.012	0.017	0.037	0.041
	(0.013)	(0.013)	(0.027)	(0.027)
2011	-0.013	0.053***	0.039	0.074***
	(0.012)	(0.012)	(0.024)	(0.025)
2014	-0.004	0.060***	0.044*	0.080***
	(0.012)	(0.013)	(0.024)	(0.025)
2017	0.040***	0.102***	0.104***	0.135***
	(0.015)	(0.015)	(0.029)	(0.030)
Constant	6.775***	6.625***	6.813***	6.709***
	(0.033)	(0.033)	(0.043)	(0.044)
Observations	15,819	15,819	5,622	5,622
R-squared	0.284	0.325	0.127	0.154

Appendix 5. Decomposition of Variances

		Male									
		Non-Union					Union				
		20	30	40	50	subtotal	20	30	40	50	subtotal
Share	2005	18%	35%	23%	24%	100%	17%	29%	26%	27%	100%
	2008	15%	36%	26%	23%	100%	16%	29%	29%	27%	100%
	2011	14%	36%	28%	22%	100%	18%	28%	28%	26%	100%
	2014	12%	32%	35%	20%	100%	17%	27%	31%	25%	100%
	2017	13%	31%	33%	23%	100%	16%	29%	32%	22%	100%
Mean	2005	7.10	7.52	7.70	7.80	7.55	7.28	7.72	8.00	8.13	7.83
	2008	7.19	7.49	7.74	7.80	7.58	7.32	7.72	8.04	8.13	7.86
	2011	7.20	7.52	7.64	7.78	7.56	7.40	7.71	7.98	8.14	7.84
	2014	7.22	7.54	7.65	7.81	7.60	7.41	7.76	7.95	8.13	7.85
	2017	7.37	7.57	7.68	7.84	7.64	7.45	7.80	7.99	8.17	7.89
Var	2005	0.258	0.219	0.244	0.258	0.298	0.223	0.162	0.141	0.202	0.264
	2008	0.237	0.177	0.226	0.232	0.254	0.280	0.167	0.217	0.212	0.292
	2011	0.221	0.180	0.216	0.310	0.256	0.196	0.192	0.159	0.194	0.252
	2014	0.294	0.203	0.233	0.290	0.271	0.276	0.176	0.206	0.193	0.265
	2017	0.227	0.200	0.241	0.373	0.276	0.250	0.157	0.204	0.228	0.257
Within	2005	0.047	0.078	0.055	0.061	0.241	0.038	0.048	0.037	0.056	0.178
	2008	0.036	0.063	0.060	0.052	0.211	0.044	0.048	0.062	0.057	0.211
	2011	0.031	0.066	0.059	0.069	0.224	0.035	0.054	0.045	0.051	0.184
	2014	0.037	0.066	0.081	0.059	0.243	0.047	0.047	0.063	0.049	0.207
	2017	0.030	0.063	0.080	0.084	0.256	0.041	0.046	0.065	0.051	0.203
Between	2005	-	-	-	-	0.057	-	-	-	-	0.087
	2008	-	-	-	-	0.044	-	-	-	-	0.081
	2011	-	-	-	-	0.031	-	-	-	-	0.068
	2014	-	-	-	-	0.029	-	-	-	-	0.058
	2017	-	-	-	-	0.021	-	-	-	-	0.055

		Female									
		Non-Union					Union				
		20	30	40	50	subtotal	20	30	40	50	subtotal
Share	2005	27%	27%	21%	24%	100%	38%	34%	12%	16%	100%
	2008	23%	33%	21%	23%	100%	39%	29%	20%	12%	100%
	2011	22%	31%	24%	23%	100%	38%	29%	20%	13%	100%
	2014	20%	29%	29%	22%	100%	36%	30%	23%	12%	100%
	2017	21%	30%	29%	19%	100%	35%	29%	21%	14%	100%
Mean	2005	7.05	7.36	7.20	7.39	7.25	7.13	7.37	7.33	7.63	7.32
	2008	7.14	7.35	7.36	7.37	7.31	7.19	7.51	7.51	7.65	7.40
	2011	7.16	7.33	7.38	7.38	7.32	7.26	7.43	7.71	7.74	7.46
	2014	7.23	7.36	7.39	7.35	7.34	7.25	7.53	7.60	7.81	7.48
	2017	7.20	7.43	7.45	7.37	7.38	7.32	7.69	7.70	7.77	7.57
Var	2005	0.303	0.214	0.330	0.187	0.273	0.305	0.279	0.359	0.230	0.317
	2008	0.152	0.175	0.291	0.176	0.202	0.255	0.211	0.305	0.184	0.272
	2011	0.224	0.236	0.306	0.207	0.250	0.299	0.297	0.168	0.175	0.293
	2014	0.250	0.231	0.251	0.280	0.254	0.277	0.266	0.317	0.246	0.313
	2017	0.192	0.234	0.254	0.150	0.222	0.287	0.251	0.297	0.397	0.325
Within	2005	0.082	0.059	0.069	0.046	0.256	0.117	0.095	0.043	0.036	0.291
	2008	0.034	0.058	0.062	0.040	0.195	0.099	0.061	0.060	0.022	0.244
	2011	0.049	0.073	0.073	0.049	0.243	0.113	0.087	0.033	0.023	0.256
	2014	0.050	0.067	0.072	0.062	0.251	0.100	0.079	0.071	0.029	0.279
	2017	0.041	0.070	0.075	0.029	0.215	0.101	0.074	0.062	0.057	0.294
Between	2005	-	-	-	-	0.019	-	-	-	-	0.030
	2008	-	-	-	-	0.008	-	-	-	-	0.031
	2011	-	-	-	-	0.007	-	-	-	-	0.038
	2014	-	-	-	-	0.003	-	-	-	-	0.036
	2017	-	-	-	-	0.009	-	-	-	-	0.036

Appendix 6. The Effects of Unions on the Regular Employment

		Male					Female				
		20	30	40	50	subtotal	20	30	40	50	subtotal
Union	2005	51%	48%	57%	57%	53%	47%	43%	26%	28%	38%
Rate	2008	54%	48%	55%	57%	53%	55%	38%	40%	27%	41%
	2011	51%	39%	46%	50%	45%	47%	33%	30%	22%	34%
	2014	55%	43%	44%	53%	47%	50%	36%	31%	23%	36%
	2017	55%	47%	48%	49%	49%	53%	40%	32%	33%	40%
□ w(c)	2005	0.18	0.20	0.30	0.33	0.28	0.08	0.01	0.13	0.24	0.07
	2008	0.13	0.23	0.30	0.33	0.28	0.05	0.16	0.15	0.28	0.09
	2011	0.20	0.19	0.34	0.36	0.28	0.10	0.10	0.33	0.36	0.14
	2014	0.19	0.22	0.30	0.32	0.25	0.02	0.17	0.21	0.46	0.14
	2017	0.08	0.23	0.31	0.33	0.25	0.12	0.26	0.25	0.40	0.19
□ v(c)	2005	-0.035	-0.057	-0.103	-0.056	-0.033	0.002	0.065	0.029	0.044	0.044
	2008	0.043	-0.009	-0.009	-0.020	0.038	0.103	0.036	0.014	0.008	0.070
	2011	-0.025	0.012	-0.057	-0.116	-0.003	0.075	0.061	-0.138	-0.032	0.043
	2014	-0.018	-0.027	-0.027	-0.097	-0.006	0.027	0.035	0.066	-0.034	0.060
	2017	0.024	-0.043	-0.038	-0.145	-0.018	0.095	0.017	0.042	0.247	0.103
Difference in Variance	2005	-	-	-	-	-0.034	-	-	-	-	0.012
	2008	-	-	-	-	0.001	-	-	-	-	0.020
	2011	-	-	-	-	-0.023	-	-	-	-	0.002
	2014	-	-	-	-	-0.021	-	-	-	-	0.010
	2017	-	-	-	-	-0.024	-	-	-	-	0.038
Difference in Average	2005	-	-	-	-	0.017	-	-	-	-	0.004
	2008	-	-	-	-	0.017	-	-	-	-	0.007
	2011	-	-	-	-	0.020	-	-	-	-	0.012
	2014	-	-	-	-	0.017	-	-	-	-	0.013
	2017	-	-	-	-	0.016	-	-	-	-	0.017
Total Effects	2005	-	-	-	-	0.005	-	-	-	-	0.018
	2008	-	-	-	-	0.042	-	-	-	-	0.030
	2011	-	-	-	-	0.012	-	-	-	-	0.018
	2014	-	-	-	-	0.007	-	-	-	-	0.028
	2017	-	-	-	-	0.010	-	-	-	-	0.058

References

Bennett, James T., and Bruce E. Kaufman. 2007. *What do unions do? : A twenty-year perspective.* Transaction Publishers.

Blanchflower, David G., and Alex Bryson. 2004. What effect do unions have on wages now and would Freeman and Medoff be surprised? *Journal of Labor Research* 25 (3): 383-414.

Blanchflower, David G., and Richard B. Freeman. 1992. Unionism in the United States and other advanced OECD countries. *Industrial Relations: A Journal of Economy and Society* 31 (1): 56-79.

Blau, Francine D., and Lawrence M. Kahn. 1996. International differences in male wage inequality: Institutions versus market forces. *Journal of Political Economy* 104 (4): 791-837.

Card, David. 1996. The effect of unions on the structure of wages: A longitudinal analysis. *Econometrica: Journal of the Econometric Society*: 957-79.

Card, David, Thomas Lemieux, and W. Craig Riddell. 2004. Unions and wage inequality. *Journal of Labor Research* 25 (4): 519-59.

DiNardo, John, Nicole M. Fortin, and Thomas Lemieux. 1996. Labor market institutions and the distribution of wages, 1973-1992: A semiparametric approach. *Econometrica* 64 (5): 1001-44.

Farber, Henry S. 1983. The determination of the union status of workers. *Econometrica* 51 (5): 1417-37.

Fortin, Nicole M., and Thomas Lemieux. 1997. Institutional changes and rising wage inequality: Is there a linkage? *Journal of Economic Perspectives* 11 (2): 75-96.

Freeman, Richard B. 1980. Unionism and the dispersion of wages. *ILR Review* 34 (1): 3-23.

———. 1982. Union wage practices and wage dispersion within establishments. *ILR Review* 36 (1): 3-21.

Freeman, R. B., and J. L. Medoff. 1984. *What do unions do?* New York: Basic Books.

Gordon, Andrew. 2017. New and enduring dual structures of employment in Japan: The rise of nonregular labor, 1980s–2010s. Social Science Japan Journal 20 (1): 9-36.

Hara, Hiromi, and Daiji Kawaguchi. 2008. The union wage effect in Japan. *Industrial Relations: A Journal of Economy and Society* 47 (4) (10/01; 2021/05): pp.569-90.

Jann, Ben. 2008. The Blinder–Oaxaca decomposition for linear regression models. *The Stata Journal* 8 (4): 453-79.

Kahn, Lawrence M. 2000. Wage inequality, collective bargaining, and relative employment from 1985 to 1994: Evidence from fifteen OECD countries. *Review of Economics and Statistics* 82 (4): 564-79.

Kambayashi, Ryo. 2013. Differences in definitions of nonregular employees in government statistics. Japan Labor Review 10 (4): 55-66.

Lemieux, Thomas. 1998. Estimating the effects of unions on wage inequality in a panel data model with comparative advantage and nonrandom selection. *Journal of Labor Economics* 16 (2): 261-91.

Lewis, H. Gregg. 1986. Union relative wage effects. *Handbook of Labor Economics* 2 : 1139-81.

Nimura, Kazuo. 1994. "Sengo-shakai no kiten ni okeru rodokumiai-undo [The Labor Union Movement at the Beginnings of Post-WWII Society in Japan]," In Watanabe Osamu et al. eds., *Nihon kingendaishi - Kôzô to Hendô 4. Sengo kaikaku to gendai shakai no keisei (The History of Modern Japan - Structure and Change, vol.4, Post-War Reforms and the Formation of Modern Society).* Iwanami Shoten (English translation: http://nimura-laborhistory.jp/English/en-postww2laborunion.html)

Nitta, Michio, and Takehisa Shinozaki. 2008. "Roudou Kumiai no Chingin Kouka." [Union Wage Effect.] In Ichiro Tanioka, Michio Nitta, and Noriko Iwai ed., *Nihon Jin no Ishiki to Koudou [The Mind and Behavior of Japanese People]*, Tokyo: University of Tokyo Press: pp. 121–33.

Noda, Tomohiko. 1997. Chingin kozo to kigyobetsu kumiai [wage structure and enterprise labor unions]. *Nihon Keizai Kenkyu*: 26-44.

Osawa, Mari. 2001. People in irregular modes of employment: Are they really not subject to discrimination? Social Science Japan Journal 4 (2): 183-99.

Suzuki, Kyoko. 2020. The Latent Structure of the Japanese Labor Market and the Type of Employment: Latent Class Analysis with Finite Mixture Model. *Japan Labor Issues* 22: 42 – 58.

Tachibanaki, Toshiaki, and Tomohiko Noda. 2000. *The economic effects of trade unions in Japan.* Springer.

Todate, Mitsunori. 2010. Economic effect of labor unions. *Japanese Economy* 37 (1): pp.111-29.

Tsuru, Tsuyoshi. 2010. "The union wage premium, voice, and nonunion workers' attitudes: Before and after Japan's lost decade." In Eriksson, T, ed., *Advances in the Economic Analysis of Participatory & Labor-Managed Firms.* Bingley, Emerald Group Publishing Limited: pp.161-204.

V. 2021 LERA Lifetime Achievement Award

DENNIS DABNEY
Kaiser Permanente

Remarks

Thanks to everyone at LERA for this thoughtful recognition—and for all the friendship and camaraderie over the years. I've come a long way since I was placed into my first labor relations job, when I was a much younger man working at a large automobile parts manufacturing plant in Detroit.

My career in labor relations led me from the automobile industry to the energy sector and, for the past nine and a half years, into healthcare, where I led labor relations for Kaiser Permanente. I've learned a lot at each stop.

Throughout my career journey, I've found LERA to be an invaluable resource in helping me to test ideas and to learn the latest industry standards from some of the best in the business. I've also made a lot of friends.

I would like to thank all of you individually by name, but there are so many of you who have helped me over the years that doing so would take all day. Suffice it to say, my association with LERA—and with all of you—has always provided me with an extra edge in my career and with an endless source of fun stories and good times. I wish each of you continued success in helping organizations succeed and employees to find fulfillment.

Thank you.

VI. The Old Gig Economy

The Extent of Payroll Fraud in Construction, Its Cost to Society, and Approaches to Its Regulation

Chair

William Canak, PhD, Professor Emeritus of Sociology, Middle Tennessee State University

Discussants

Matthew F. Capece, Esq., Representative of the General President, United Brotherhood of Carpenters & Joiners of America (UBC)

Katherine G. Abraham, PhD, University of Maryland

Panelists

Mark Erlich, Fellow, Labor and Worklife Program, Harvard University Law School, Retired, Executive Secretary Treasurer, New England Regional Council of Carpenters

Dale Belman, PhD, Professor, Michigan State University, School of Human Resources and Labor Relations, President, Institute for Construction Economics Research

Rebecca Smith, Esq., National Employment Law Project

David Weil, PhD, Dean, Heller School for Social Policy and Management, Brandeis University, Former Administrator Wage & Hour Division, US Department of Labor

Discussant Comments, Complementary Data, and Research Opportunities

Introduction and Perspective

My work with the carpenters' union since 1989 has focused on improving enforcement of labor standards and employment tax laws in the troubled construction industry. That charge has required accumulating evidence of violations, interacting with law enforcement, plaintiff counsel, contractors, researchers, community organizations and other stakeholders, the media, and legislators. The work requires collaborating with others not only to create policy but to see it put to the test. That experience informs my comments, many of which address additional research to complement the work of the panelists.

Summary of Presentations

The panelists' presentations covered the existence, history, degree, trends, and factors contributing to fraudulent employment practices in the construction industry, commonly referred to as misclassification as independent contractors or "off-the-books" (unreported) employment.

Mark Erlich's experience as Executive Secretary Treasurer of the New England Council of Carpenters provided a perspective coming from an industry stakeholder who experienced the milestones leading up to the current troublesome state of affairs in the construction industry. The diminishment of labor and employment law enforcement and union representation, Erlich detailed, has given a competitive advantage to employers that violate employment laws, allowing them to dominate construction markets. His paper, *Misclassification in Construction, The Original Gig Economy*[1] is a must-read on the topic.

Another must-read is the first comprehensive national study quantifying the losses caused by fraudulent employment practices in the construction industry, *An Empirical Methodology to Estimate the Incidences and Costs of Payroll Fraud in the Construction Industry*, which was coauthored by Dale Belman, Mark Erlich, and Russell Ormiston.[2] The paper was presented by panelist Belman. As the title, suggests, the paper details a recommended methodology for quantifying lost federal and state income taxes, employment taxes, workers' compensation premiums and workers' wages. There have been numerous studies done in states, with varying methodologies. The paper's methodology section is a significant contribution because it invites standardization for future research. Their methodology, which they characterize as the "indirect method," compares government data estimating total employment to legal employment.[3]

The numbers resulting from their research are staggering. The authors concluded that a yearly average of 1.3 to 2.16 million construction workers, or 12.4 to 20.5 percent of the construction workforce, are either misclassified as independent contractors or paid off the books.[4] Workers lose close to $1 billion in overtime and other premium pay. Using more aggressive assumptions, employers operating illegally reduce their labor costs by $17.33 billion.[5]

In the background is the brewing threat of the independent-contractor operation models of Uber, Lyft, DoorDash, and others in the so-called gig industry.[6] Rebecca Smith of the National Employment Law Project reviewed the gig industry's record on substandard employment conditions and their efforts to reshape employment law—to create a third category of employment to legitimize their business model. Of particular concern has been the gig industry's success in passing Proposition 22 in California and their plans to institute similar proposals in other states. Some states have already created laws that except the gig industry from their employment laws.

David Weil recognized the exploitative employment practices in the construction industry and the contributing role of fissuring. He presented a paper coauthored with Tanya Goldman, *Who's Responsible Here? Establishing Legal Responsibility in the Fissured Workplace.*[7] The authors recognize changes in the nature of work due to fissuring, the lack of meaningful law enforcement (as we see in the construction industry), and courts narrowing the reach of employment protections.[8] Their recommended solution is to adopt a new framework of three concentric circles of rights and responsibilities.[9] In the middle are fundamental rights that all workers have no matter their classification: freedom from retaliation and discrimination, safety, appropriate compensation, "and freedom of association and the right to engage in acts for mutual aid and protection."[10] The middle circle includes a presumption of employment using the ABC test. Workers who are employees would enjoy all of the protections that come with that status in addition to the fundamental protections. Those fundamental protections would continue to apply to independent contractors.[11] The outer circle includes access to social-safety-net protections: workers' compensation, unemployment insurance, paid family and medical leave, retirement and skill training, and nonmandatory benefits for employees and independent contractors alike. Benefits would be portable and paid for by employers for employees and self-paid by independent contractors.[12]

A Recommended Additional Data Source

The following sentences stand out from the Ormiston, Belman, and Erlich study:

> In developing the cost estimates, ... the authors have used conservative assumptions whenever possible. This includes, but is not limited to, considering only the most conservative number of workers directly affected (1.3 million) in the ranges

> presented. … However, the authors suspect—even if they cannot verify—that the social costs of payroll fraud may be substantially larger."[13]

The absence of more precise data requires analysts to apply assumptions, and often conservative assumptions, to preserve credibility. Additional data sources would indeed assist this perplexing and persistent need for more precision. An opportunity for more precision can be scored, with some coaxing of the US Department of Labor Employment and Training Administration (ETA) and state unemployment divisions.[14]

The ETA has performance measures for state unemployment insurance audit divisions.[15] State audit divisions are required to file Form 581 with ETA, which details, among other items, the number of employees found to have been misclassified as independent contractors. Data from the form is made available to the public in a report organized per region and state.[16] ETA Form 581 can be modified to collect information on the number of misclassified workers per industry. If would further be helpful if the data included misclassified workers found by random and targeted audits. This would assist state auditors and federal regulators by disclosing troublesome industries on which to focus enforcement initiatives. It additionally creates another data point for researchers examining specific industries. Currently, getting such data would require researchers to make an extraordinary request to state unemployment divisions.[17]

There may be other opportunities to get additional data sources through wrench turning with agencies and regulations. Some focus on this issue in future presentations could prove to be helpful for stakeholders and ultimately to economists and other researchers.

More Data Are Needed on Enforcement Capabilities

Below are quotes from a news article on the findings of West Virginia legislative auditors, who analyzed the state unemployment insurance agency's ability to collect revenue:[18]

> In a yearlong examination of WorkForce West Virginia, legislative auditors found that employee misclassification was "rampant," leading to a loss of $824,000 to $1.1 million in personal income tax revenue from 2014 to 2018.
>
> During each of those years, the US Department of Labor gave West Virginia a "fail" rating for its efforts in detecting employee misclassification.
>
> Part of the problem, according to the audit, is [that] WorkForce lacks its own auditors. The agency employed one full-time auditor in 2019, compared to seven in 2014. Low pay hinders recruitment.
>
> Employee misclassification isn't unique to West Virginia, one of 26 states and territories that received the "fail" rating in 2014 and 19 that received it in 2017.
>
> Auditors said a posting for a tax examiner required candidates to have a bachelor's degree with 24 hours in accounting for a job with a starting annual salary of $27,729. The requirements were similar for a tax and revenue auditor, with a starting salary of $31,146, according to the report.

Compare West Virginia's lone auditor to Florida's enforcement of workers' compensation coverage requirements in the construction industry. Florida requires, with some exception, employees and independent contractors in the construction industry to be covered with workers' compensation insurance.[19] Its enforcement laws have robust administrative fines, stop-work orders, and felony criminal penalties.[20]

Additionally, compared to other states, Florida's is well-staffed with auditors, insurance-fraud detectives, and even dedicated workers' compensation fraud county prosecutors.[21] Still, it is well known to us in the carpenters' union that misclassification and off-the-books employment is the rule rather than the exception on construction projects surveyed in the state.[22] Why is that, when laws are strong and there appears to be sufficient personnel?

The prevalent practice is to analyze our employment and tax laws and recommend repairs, especially with defining "employment." That is valuable, but shortcomings in statutes and judicial interpretations are not the only reason working conditions for many in construction and other industries have deteriorated. Agency budgets, leadership, technology, caseloads, strategy, and turnover have a tremendous impact on enforcement outcomes. The best law will have little to no meaning when enforcement is left to only one low-paid auditor or a revolving door of overwhelmed prosecutors.

Certainly, law enforcement and advocate communities would benefit from additional analysis of the capabilities of state and federal agencies that are entrusted with brining wrong-doers to justice. There has been some. Although it posed more questions than answers, one example is the study by the Columbia Law School National State Attorney's General Program on state wage and hour enforcement in 2011.[23] From the research and other good work done by the panelists, we know that outcomes demonstrate current capabilities are not meeting the challenge. Pinpointing the capability problems and exposing them with data can inform advocates and lead to solutions.

Employer Losses Have Been Recognized but Need More Attention

Construction workers, taxpayers, and workers' compensation insurers are not the only ones suffering from the epidemic of wage theft and fraud in the construction industry. The construction market is hypercompetitive. Contractors who adopt law-breaking schemes can shave their labor costs by 16.7 to 48.1 percent.[24] As the panelists recognized, the resulting lower bids allow scofflaws to steal work away from law-abiding employers.[25] Thus, the degradation of the industry grows. Law-abiding construction employers who do not want to take the low road know this and are not pleased.[26]

A thorough analysis of the profits lost by law-abiding contractors and the additional insurance[27] and taxes they pay would provide powerful and useful information for policy makers and fill a gap in the literature.

Endnotes

[1]Mark Erlich, Misclassification in Construction, The Original Gig Economy, *ILR Review (*November 26, 2020). https://bit.ly/3ehPFMu

[2]Russell Ormiston, Dale Belman, and Mark Erlich, An Empirical Methodology to Estimate the Incidences and Costs of Payroll Fraud in the Construction Industry (January 2020). The United Brotherhood of Carpenters commissioned the research. https://bit.ly/3ejUiWe

[3]*Id* at 3 and 20. Since the panel presentation, a paper quantifying fraudulent employment practices in Wisconsin, Minnesota and Illinois was released following their recommended indirect method. Nathaniel Goodell and Frank Manzo IV, The Costs of Wage Theft and Payroll Fraud in the Construction Industries of Wisconsin, Minnesota, and Illinois: Impacts on Workers and Taxpayers, 2, 4 and 6 (January 14, 2021). https://bit.ly/3FlDA4B

[4]Ormiston, *supra* note 2, at 5, Table A.

[5]*Id* at 51.

[6]The National Employment Project has a paper on this topic focusing on the Proposition 22 campaign in California: Rey Fuentes, Rebecca Smith and Brian Chen, Rigging the Gig: How Uber, Lyft and DoorDash's Ballot Initiative Would

Put Corporations Above the Law and Steal Wages, Benefits, and Protections from California Workers. Partnership for Working Families and the National Employment Law Project (July 2020). https://bit.ly/3mr6TM2

[7]Tanya Goldman and David Weil, Who's Responsible Here? Establishing Legal Responsibility in the Fissured Workplace, *Berkeley Journal of Employment & Labor Law*, Vol 42:1.

[8]*Id* at 27 and 51-52.

[9]*Id* at 56.

[10]*Id* at 56-57.

[11]*Id* at 69-71 and 79.

[12]*Id* at 81-83.

[13]Ormiston, *supra* note 2 at 6.

[14]Formal rulemaking under the Administrative Procedures Act may also be needed.

[15]Jane Oates, Assistant Secretary, Employment and Training Administration Advisory System, US Department of Labor, Unemployment Insurance Program Letter NO. 03-11, Implementation of Effective Audit Measure (December 30, 2010). https://bit.ly/3GWMmXo

[16]See, e.g., Education and Training Administration, US Department of Labor, Average Number of Misclassified Employers Detected per Audit Calendar Year Ending December 31, 2019. https://bit.ly/3sp2EnS

[17]Getting specific industry data from state audit unemployment audit divisions is very difficult but not impossible. Dale Belman and Richard Block were successful in getting such data for a report they wrote on employee misclassification in Michigan. Dale Belman and Richard Block, Informing the Debate: The Social and Economic Costs of Employee Misclassification in Michigan, Michigan State University, Institute for Public Policy and Social Research, 7 (2008). https://bit.ly/30PZH4f

[18]Lacie Pierson, Audit finds West Virginia could be missing $820K to $1.1M in income tax revenue, *Charleston Gazette-Mail*, December 28, 2020. https://bit.ly/3EfYZuZ. I am uncertain about whether or not the legislature has given WorkForce additional resources to do their job, but I do know that they have just passed a bill signed into law by the governor that makes more workers in the state independent contractors, so they will be ineligible for unemployment, workers' compensation, wage, and anti-discrimination protections. See S.B. Sub. 272, 2021 Sess. (W.V. 2021).

[19]Fla. Stat. § 440.02(15)(b)(2), (c)(2), (3) and (4) (2021).

[20]Fla. Stat. §§ 440.105, 440.107, 440.381 (2021).

[21]See, e.g., Florida Department of Financial Services, Division of Investigative and Forensic Services Bureau of Workers' Compensation Fraud and Division of Workers' Compensation, Joint Report to the President of the Florida Senate, the Speaker of the Florida House of Representatives (January 15, 2019). https://bit.ly/3FoeVg3

[22]The construction industry in Florida is also used by organized crime to launder money. David Borum and Geoffrey Branch, How construction cons steal workers-comp premiums: It's a shell game, *Journal of Insurance Fraud in America*, February 2017, reprinted by NU Property Casualty360. https://bit.ly/3yPmutu

[23]Jacob Meyer and Robert Greenleaf, Enforcement of State Wage and Hour Laws: A Survey of State Regulators, Columbia Law School National State Attorneys General Program (April 2011). https://bit.ly/3qBVASv.There has also been reporting on the decimation of IRS enforcement by ProPublica. Jesse Eisinger and Paul Kiel, After Budget cuts, the IRS' Work Against Tax Cheats Is Facing "Collapse," ProPublica, October 1, 2018. https://bit.ly/3ekaSoW.

[24] Karl A. Racine, Attorney General for the District of Columbia, economic analysis by Dale Belman and Aaron Sojourner, Illegal Worker Misclassification: Payroll Fraud in the District's Construction Industry, 1, 2 and 15 (May 22, 2019). https://bit.ly/3ei6C9A.

[25]Ormiston, *supra* note 2 at 2.

[26]See, e.g., Hearing on Misclassification of Employees: Examining the Costs to Workers, Businesses, and the Economy: Before the Workforce Protections Subcommittee, House Education and Labor Committee, 116th Congress (2019) (Statement of Matt Townsend, President of the Signatory Wall and Ceiling Contractors Alliance), 2, 3 and 6. https://bit.ly/3piN4bt; and Doug Burton, op-ed, To help NC businesses, end the misclassification fraud, *The News & Observer*, June 3, 2015. https://bit.ly/3swSD86.

[27]We know workers' compensation insurers pass along their premium-fraud losses to insureds that actually pay. There has been some research on the extra premium cost for law-abiding construction employers in California. Frank Neuhauser and Colleen Donovan, Fraud in Workers' Compensation Payroll Reporting: How Much Employer Fraud

Exists and How are Honest Employers Impacted? University of California, Berkeley, 31 (August 2007) ("[H]onest employers are paying substantially more, up to eight times more, for workers' compensation than actual experience would imply is accurate. This is a substantial transfer of income and profits from honest employer [sic] to dishonest employers.") Of course, when workers' compensation costs increase for good construction employers it makes them even less competitive against the scofflaws.

VII. LERA Annual Reports

LERA Executive Board Meeting Minutes
Friday, February 5, 2021, 12 noon Eastern Time
Via Videoconference

Call to order—The meeting was called to order at 12:02 p.m. by President Adrienne Eaton. Present at the meeting were Dan Altchek, William Canak, Bonnie Prouty Castrey, Joel Cutcher-Gershenfeld, Robert Chiaravalli, Paul Clark, Joel Cutcher-Gershenfeld, Dennis Dabney, Bill Dirksen, Adrienne Eaton, Julie Farb, Cyndi Furseth, Shannon Gleeson, Kate Griffith, Erin Johansson, Harry Katz, Ryan Lamare, Tamara Lee, Kevin Legel, David Lewin, Wilma Liebman, Brad Markell, Jim Pruitt, Javier Ramirez, William Rodgers, Jake Rosenfeld, Emily Smith, Aaron Sojourner, Maite Tapia, Bernadette Tiemann, Andrew Weaver, and David Weil.

Approval of the Minutes—The motion, seconded by Bill Canak, to approve the minutes passed unanimously from the board meeting in June 2020 was approved unanimously. The motion to approve the minutes from the general membership meeting in June 2020 also was approved unanimously.

Reports

Financial Report—Andrew Weaver reports that the financials for 2020, despite the situation of the pandemic, were strong. There was an expected and significant hit to meeting revenue, but expenses were also decreased correspondingly so that the net margin remained in the black. In addition to controls on expenses, individual donations and sponsorships held steady and assisted in plugging gaps. Our net margin in 2020 was smaller than in 2018 or 2019, not surprising given the challenges faced, but this does represent the sixth year in a row of a sustainable business model, after years of consecutive losses. Looking forward, we have seen some softness in the initial membership numbers in 2021, and therefore the new budget prudently reduces the expected membership revenue. Likewise, the membership outreach and promotion budget has been increased. As we will meet again virtually in 2021, the annual expenses and revenues for the meetings category are likewise reduce, but it remains an important category to realize revenue. The 2021 budget was approved by the board unanimously.

LERA 73rd Annual Meeting Program Committee Report—Program Chair and President-Elect Wilma Liebman thanked vice chairs Aaron Sojourner and Lynne Rhinehart. The Program Committee entered into the process expecting to meet in person in Detroit, Michigan, but, of course, those plans have changed and the June 2021 meeting will take place virtually. A Detroit sub-group had been established, headed by Bob Chiaravalli, but they will hold until next year, when we plan to meet in Detroit in June 2022.

To give you a quick overview of the planned program, there were many submissions along the lines of the theme of the program, the question of if this is a transformational moment for work and worker power in the workplace in an era of division and disruption. We decided that given the intersecting crises that we as a nation are facing, that we wanted to have four plenary sessions.

The first plenary will take place on Saturday, the opening day of the conference on the topic of the Biden-Harris administration's labor and employment policy. Our speaker has yet to be determined. The second plenary will take place Saturday afternoon, and it's entitled "The COVID-19 Crisis Shaping the Future of Work", moderated by Dennis Dabney and feature speakers David Autor, Massachusetts Institute of Technology; Kimberly A. Lawrence, CVS Health Corporation; and Rebecca Dixon, National Employment Law Project. The third plenary will be Monday morning, "Meeting the Challenge of Racial Equity and Justice in the Workplace, moderated by Fred Alvarez, Coblentz Patch Duffy & Bass and Former EEOC Commissioner; and Jenny Yang, Former Chair, EEOC, and featuring Darrick Hamilton, The New School; Fred Redmond, United Steelworkers; and Johnna Torsone (invited), Pitney Bowes. The fourth plenary will take place Tuesday morning

"Competing Ideas on a Way Forward for America's Workers", moderated by Cynthia Estlund, New York University, and featuring Oren Cass, American Compass; and Damon Silvers, AFL-CIO.

In addition to the plenaries, we have a wide range of sessions: roundtables, panels, symposia, and workshops that deal with issues in the workplace from a variety of perspectives, and there will be quite a few international speakers, likely arising from the virtual nature of the meeting. The meeting will have good coverage of international and comparative topics, particularly looking at how COVID-19 was managed in the workplace. There will be coverage of virtual dispute resolution, and a number of sessions on racial justice including police reform, examining racial equality and inequality in the American South, critical race theory and industrial relations. We also have a number of sessions dealing with technology and how unions are dealing with technology and surveillance issues. We also had a notable proposal on the crisis at sea examining how seafarers fared during COVID-19 with several international speakers from various international organizations. We also have a session on Native American workers and how they are faring in terms of economic opportunity. And, of course, we have a wide range of "LERA-type" issues on themes and issues of general interest to both practitioners and scholars. I would have liked to have seen even more practitioner participation, and tried to get it, but overall I believe we have a good sampling.

A full discussion about the fees registration at the virtual conference covered the following points. Last year, our webinars did bring in some new members, and perhaps this should be considered when pricing the virtual conference for 2021. Should we ask speakers to pay for the whole conference when they only wish to attend one session? LERA's operating budget was discussed, as was a potential one-day rate for the virtual conference. In years of place-based meetings, our regular early bird member registration rate is $335, and our one-day conference registration fee is $170. Discounting the full virtual conference registration rate to $150 for members and non-members alike is less than our one-day member rate in typical years, and less by more than half of our early bird full-conference member rate. Our mission is to allow as many people to attend as possible and hear from multiple perspectives by keeping our registration as low as possible and asking for people to pay their own way. Support was heard for keeping the conference registration fee as proposed because LERA is the one place you can attend and see where the entire field is heading, and also because it is a way of supporting the association. In the past, we have had some very lean years, and many discussions about whether or not LERA would survive financially. It was mentioned that organizations and companies could be encouraged to support registration for those that should be encouraged to attend, and that we have also now established a fund to support meeting registration for those who cannot otherwise afford to attend and meet our goal of increasing diversity and inclusion at the meeting. From a university-perspective, LERA should consider that many universities and other organizations are experiencing financial difficulties and are struggling, putting all types of non-mission critical activities in question, such as traveling to and attending conferences like LERA's. Questions were raised about the type of funding that exists to offset registration fees for people from under-represented minority groups, and some discussion was had about the Diversity and Inclusion committee, which oversees this process. Last year, we accepted applications from session organizers for people that they felt were important to include within their sessions that needed financial assistance, and all of those applications were accepted in 2020. This year, we will utilize a brief form through which organizers can submit a request. We do plan to work with our organizational members as well in 2021, to assist students and apprentices to be able to join the conference affordably. Secretary-Treasurer Andrew Weaver offered an organizational financial perspective that our budget for 2021 does not have much cushion and decreasing meeting registration for several hundred participants would have a significant impact on our overall budget for the year. Many of our expenses are fixed and he encouraged the board to utilize tools like the Diversity and Inclusion fund to encourage participation where needed, but to leave the registration schedule unchanged for the 2021 virtual annual meeting.

Membership Committee Report—Cyndi Furseth, board member and Membership Committee Co-Chair, reported that the historical membership trend, as communicated in attachment D, has rebounded in recent years, which is excellent news, and the committee largely attributes this to efforts conducted by the staff. We did pick up some new members last year with our COVID-19 webinars. The committee feels that the member dues are reasonable and plans no changes in 2021. We are 52% academic in our demographics, and we do have work to do in terms of reaching out to practitioners and meeting their needs. We would like to engage more unions. They have great people who could benefit the organization, and I'm not sure how many of them know about LERA. Something we have noticed is that we have great engagement around our big events, and then a lag, so we were happy to see events like our virtual series of webinars to keep people engaged in-between. I have personally interfaced with several management prospects and asked them to become members, in union and non-union settings, and they have all agreed, but it appears we have even more work to do to engage management members, as our percentage is in decline. I encourage each of us to reach out with personal contact and phone calls to those that should join the organization.

David Lewin, Membership Committee Co-Chair, reported that in the last decade participation has increased with academics and decreased with management and attorneys. One suggestion to increase management membership in LERA is to reach out to the California Employment Law Council, which among other things deals with labor and employment matters. Due to changes in some California industries, there is a movement away from mandatory arbitration of employment disputes for some. There are unionization initiatives underway at several high-tech companies, and individuals at the California Employment Law Council as well as staff from the companies involved may be interested in the kind of research and its practical implications that LERA specializes in and that is featured in the program put together by Wilma Liebman and the program committee. I believe the time is right to communicate with the management side and see what progress we could make towards engaging them.

In reviewing the membership data, I also see that membership increased from 2016-2020, but it decreased between 2010 to 2016, and I'm not sure if we can draw any affect on the various federal administrations of those eras, but it is interesting to consider.

Bonnie Castrey, NCAC Vice Chair, offered that this affect may be related to administration within LERA, as she recalls her presidential year in 2016 when the primary objective was to save the organization in multiple ways, the rapid rate of decline in membership being one of them. A Strategic Thinking Committee was formed (comprised of members like Bill Canak, David Lewin, Marlene Heyser, Bonnie Castrey, and others), to create our Regional Vice Presidents and take other measures that would bring national LERA closer with our chapter members and thereby closer to practitioner members. That work continues as the NCAC works closely with our RVPs, and Bonie feels LERA could do more on this that would continue to bring our chapter and national level members together.

Robert Chiaravalli, Mid Region RVP, discussed that Bruce Kaufman authored a book "The Origins and Evolution of the Field of Industrial Relations in the United States" and that this book dedicated an entire chapter to professional associations, and he suggests that we review it. The book, in 1990, predicted the decline of professional associations like ours and even had a few suggestions for protecting our membership levels and even growing them. We do need to increase our management and union participation; we need to get more employers and more labor organizations coming to our meetings. When we convened a group of Detroit locals to support the LERA annual meeting taking place there, we created a list of 12 or 13 affinity groups, and Robert would be happy to share that with the membership committee.

Shannon Gleeson, board member, commented that since labor and employment relations is an interdisciplinary field, most of our members have multiple professional organizations from which to choose, so she sees an increase in participation from academics as a very positive sign. Our dual strategies of maintaining a practitioner presence while also maintaining relevance in the academic field may be in competition with each other at times, but she views it as a very good sign that the academic piece is remaining strong.

Brad Markell, board member, voiced support for David Lewin's outreach plan and considered it a ripe opportunity.

President Adrienne Eaton agreed and said that one way forward may be to create a series of task forces, focused on increasing management, labor, and government membership, to work in conjunction with the membership committee. This idea needs to be discussed and considered, especially the membership dues structure. Some so-called "alt-labor" groups are resource-poor, and even traditional labor groups have restricted finances. So, many of these groups do not have the funding to pay for memberships, so if we want to attract them we will need to consider the membership structure. As president, Adrienne Eaton pledged to consider this, consult with others, review who had volunteered for these task forces (Robert Chiaravalli, Paul Clark, others?) and perhaps have more concrete recommendations. Paul Clark had also mentioned a decline in government participation, and anecdotally, there does seem to be good involvement from FMCS, but not from the NLRB.

Development Committee Report—Harry Katz, Development Committee Co-Chair, reports that 2020 was a strong year in terms of development and contributions, and our unrestricted giving remained very level to that of the previous year. The organization received $36,276 in personal contributions. This is a meaningful contribution to our revenues but not an overwhelming one. We have also launched a quasi-endowment campaign, in part prompted by some very generous endowment gifts that were made to us, and you may have seen correspondence sent out encouraging others to consider LERA in their estate. This campaign has been successful; the second-half of our initial endowment gift was received in January 2020.

Organizational membership is increasing each year, and Jjm Pruitt and Harry Katz would both like to personally support the efforts of the membership committee in increasing Organizational Memberships, with personal contact. If the membership committee or others can point them in the right direction, they are happy to do the personal contact work, for both membership and sustaining gifts.

Lastly, Harry Katz discussed major sustaining sponsorships from Ford and Kaiser Permanente. Dennis Dabney deserves our thanks for all he has done to assist us over the years. 2020 again saw sustaining sponsorships from both Ford/UAW and Kaiser Permanente, for which we are very grateful.

Jim Pruitt, Development Committee Co-Chair and West Regional Vice President, remarked that Kaiser Permanente, led by Dennis Dabney, gets a great deal of credit for the last several years of contributions. Dennis has worked to secure those contributions for LERA, with Jim's help. Dennis Dabney got deeply involved with the board, and as President last year, and he deserves a great deal of our gratitude. Dennis Dabney plans to retire in 2021, and Jim pledges to continue in his wake to continue that stream of contributions.

Jim Pruitt called upon all his fellow board members to consider giving $1,000 in personal contributions to LERA, as it can be afforded on a personal basis. Jim remarked that this level of giving is available to him at his age and stage in life, but anything you can do to contribute is greatly appreciated. He underscored his availability to call an organization and make a pitch, especially labor organizations, and thanked Harry for his contributions to the committee.

When board members inquired about where their personal contributions should be directed, towards restricted or unrestricted giving, our opinion is that unless the gift is on the level of an estate gift, the please direct annual giving to the annual fund, Gershenfeld fund, or the Dunlop fund. The Gershenfeld fund was recently repurposed to support attendance at the meeting for under-represented minorities who cannot afford the cost of registration, and the Dunlop fund supports public policy discussion within LERA.

ILERA and LERA 2024 Report—Harry Katz, past LERA President and current Development Committee Co-Chair, is the incoming president for ILERA, which conducts a World Congress every three years. His term as ILERA president will begin July 2021, and one of his tasks will be to organize and arrange the World Congress in 2024. Katz reminded the board that LERA plans to conduct its 2024 meeting jointly in downtown Los Angeles with the ILERA sometime in the summer. This collaboration has been endorsed by previous boards, and though the pandemic has put off some of the details of planning, it will go forward in the coming year. Los Angeles was arrived at as the destination after some polling; it's convenient place to gather for those coming from Asia and other parts of the world, there is a great deal of union activity taking place in downtown LA, and there is general interest by internationals to visit downtown LA. Dennis Dabney kindly pointed us to a union hotel, the InterContinental, in downtown LA, and when the pandemic lifts, we will visit the property and set a contract. Please keep summer 2024 in LA in mind for your long-term planning.

LERA@ASSA Program Committee Report—Bill Rodgers, Program Committee Chair, reported that the LERA@ASSA 2021 meeting last month was an excellent showing of 18 well-attended virtual sessions, with the consensus being that perhaps the virtual nature of the conference made it easier for people to attend. Our median attendance was up from 23 to 30. The largest session conducted had 81 attendees, which is no surprise as the topic was how the pandemic affected the labor market.

The committee has an open call for proposals for the 2022 meeting out, with a March 15, 2021 deadline for submission. Please circulate the call to your colleagues.

Bill will be adding some individuals to the committee, and is in the process of naming the second co-chair. Bill had invited Erica Groshen, who respectfully declined due to a full schedule, and he has now extended an invitation to Abigail Wozniak of the Minneapolis Federal Reserve. He requested that anyone who knows her personally contact her to answer any questions she might have about the role.

The LERA@ASSA 2022 meeting will take place, hopefully in person, in Boston, Massachusetts.

Nominating Committee Report—Adrienne Eaton, LERA President, reported the recent additions to the LERA Executive Board from the last election. Paul Clark is our next President Elect. Peter Berg will represent the academic perspective. Janet Gilman will represent neutrals. Deborah Mueller, labor, and Quentin Herbert and Kevin Legel will be representing management. Jim Pruitt has been re-elected in his position as West Regional Vice President.

Adrienne Eaton thanked the nominating committee folks for the work that went into the process and for diversifying our board representation.

Editorial Committee Report—Ryan Lamare, Editor-in-Chief, reported that the 2020 research volume, edited by Dionne Pohler on "Re-imagining the Governance of Work and Employment" was successfully published and distributed in the Fall of 2020.

The 2021 Research Volume is calling "Revaluing Workers Toward Democratic and Sustainable Future" edited by Tobias Schulze-Clevin and Sean Rodgers. 13 chapters were solicited and 10 have been completed and are at various stages with the LERA copy-editing team, and we have three outstanding chapters. The book is on pace to be completed successfully in 2021.

The editorial committee has received a proposal for the LERA 2022 Research Volume, currently titled "Industrial Relations and Racial Reckoning" co-edited by Sheri Davis, Tamara Lee, Maite Tapia, and I. Williams. Tamara Lee offered further description of the volume. She mentioned that the co-editors are attempting an innovative approach, focusing on a wide range of story-tellers, ranging from poets, some of them activists, to more academic-themed work, and there will be some academic heavy hitters as well.

The motion to approve the proposal for the LERA 2022 research proposal was accepted unanimously by the board.

National Chapter Advisory Council Report—William Canak, NCAC Chair, reported that despite challenges in 2020, the chapters have remained active and done innovative work, including meeting virtually. We don't have feedback yet on the full scope of impact of the pandemic on chapter membership, but we'll be speaking to the full National Chapter Advisory Council and Regional Vice Presidents scheduled for February 10, 2021. We have a few new chapters in formation, one in Virginia, and the other in the area of Louisiana/Mississippi. It's terrific to get more chapters in the south and southeast regions. Sarah Espinosa Miller is leading the chapter building in Virginia, and Javier Ramirez of the FMCS is assisting with the Louisiana/Mississippi initiative. At the LERA 73rd Annual Meeting in June 2021, the NCAC is again planning a full complement of chapter activities including: a chapter administration workshop featuring the innovative and successful programs of New Jersey LERA (President Jonathan Cohen) and NorCal LERA (Board Member Jim Pruitt and Treasurer Renee Mayne), and a meeting of all Chapter Representatives, which last year we offered a rebate for Chapter Presidents to attend the LERA Annual Meeting if they attended the Chapter Representatives meeting, and we hope we can extend the same offer in 2021 as it did assist with attendance and representation of the various LERA Chapters at the LERA annual meeting. Our three Regional Vice Presidents (Beverly Harrison, Jim Pruitt, and Robert Chiaravalli) continue to work within their regions, and Jim Pruitt was just re-elected to the board to serve a second term in this capacity. As a reminder, since our Bylaws update in 2016, all national LERA members are required to become a member of a local chapter if there is one available to them, so please look up the chapter in your region and become a member if you haven't already done so. Bonnie Castrey, NCAC Vice Chair, reported that since that Bylaws revision, we have become much more closely connected to our LERA Chapters, which was the outcome we were hoping to achieve. With LERA being able to distribute our bi-weekly electronic newsletters to Chapter affiliates, we have become a much more cohesive organization. I'm not sure if we have the ability to track and report the affect of this communication on membership, but anecdotally speaking, we have heard reports from chapters (including my home chapter, SoCal LERA, formerly OC LERA) that chapter members are becoming full national members. Additionally, there is now a Virtual LERA chapter to join, though almost all of our chapters are now offering some level of virtual programming. Paul Clark from Penn State (the home of VLERA), reports that VLERA is flourishing. They have begun charging membership dues of $20 a year, and programs are quite successful in terms of both engagement and quality. When the program was first discussed by the board several years ago, there was some trepidation that having a virtual chapter could pull members from other local chapters, but we have not found this to impact the membership of other local chapters, and in fact, VLERA has found ways in which to support other local LERA Chapters. The innovative chapter has found its niche, and is bringing individuals to LERA that might not have otherwise known about the organization.

New Business

David Lewin, Membership Committee Co-Chair, reported on a proposal to create a high school essay competition on labor and employment relations topics to create additional interested in younger generations about labor issues. The competition could carry a prize. Do we have a paper competition specifically for undergraduate students? If we want to increase LERA's profile with high school students, we could work through the teachers' unions. We may wish to review the Best Dissertation Award (Shannon Gleeson) and the UCIRHRP Best Student Paper and Best Student Poster awards. Bonnie Castrey is willing to work on high school initiative and Julie Farb is interested in the college level competition. Paul Clark offered that it would be appropriate for the UCIRHRP to have a strong role in that initiative.

Adrienne Eaton, President, would like to develop an Interest Section for those specifically interested in education in labor relations, and then work towards organizing either an annual or bi-annual event targeted to educators in labor and employment relations. The number of non-tenure track educators is increasing, and they would like opportunities to grow and develop, and, of course, tenure-track professors are also interested in teaching. It would not be a heavy lift to get a new Interest Section started, and Joel Cutcher-Gershenfeld committed to producing a draft charter for the fledgling Interest Section. Paul Clark suggested that this is a good initiative for the UCIRHRP to get involved with and support; he pledged to assist.

William Canak, NCAC Vice Chair, would like to propose the idea of repeating our 60th Anniversary Cartoon Book for the LERA 75th anniversary. The board is encouraged to find someone at *The New Yorker* who could confirm if the company is still printing this type of book.

Adjournment—The meeting was adjourned at 2:01 pm Eastern Time by President Adrienne Eaton.

LERA Executive Board Meeting Minutes
Friday, June 1, 2021, 6 p.m. Eastern Time
Via Videoconference

Call to order—The meeting was called to order at 6:01 p.m. by Adrienne Eaton, President. Present at the meeting were William Canak, Bonnie Prouty Castrey, Robert Chiaravalli, Paul Clark, Joel Cutcher-Gershenfeld, Dennis Dabney, Virginia Doellgast, Adrienne Eaton, Julie Farb, Shannon Gleeson, Kate Griffith, Beverly Harrison, Erin Johannson, Harry Katz, Ryan Lamare, Kevin Legel, Tami Lee, David Lewin, Wilma Liebman, Jim Pruitt, Javier Ramirez, Jake Rosenfeld, Emily Smith, Matie Carmen Tapia, Bernadette Tiemann, and Andrew Weaver.

Approval of the Minutes—The motion to approve the minutes passed unanimously from the board meeting in February 2021. Paul Clark motioned to approve. Naomi Williams name should be spelled out correctly in the minutes; and with that revision, the motion passed unanimously, seconded by Bill Canak.

Reports

Executive Committee Report—Adrienne Eaton welcomed the board and thanked Wilma Liebman and the program committee for putting together the LERA 73rd Annual Meeting, and securing the Secretary of Labor as a keynote. Additionally, she thanked Paul Clark for organizing the Teaching Interest Section, and there will be more to come in the meeting later on.

Financial Report—LERA Secretary-Treasurer Andrew Weaver reports that 2020 has been a challenging time. Even though revenue is down, expenses have been held down as well. The organization is still in the black and showed a positive net income of around 6% of revenue; quite respectable performance. There has been some softening of membership dues in 2021, and we have increased the membership promotion budget accordingly to shore up the membership numbers by year-end. All things considered and from a balance sheet perspective, and our current audit, we have reasonable net assets and good cash flow in 2020. Wilma Liebman motioned to accept the financial report and Jim Pruitt seconded it, and the motion passed unanimously.

Editorial Committee Report—Ryan Lamare reports on the 2021 Research Volume "Revaluing Workers" by Tobias Schulze-Cleven and Todd Vachon is finishing up production. The 2022 Research Volume was approved at the last board meeting "Industrial Relations and a Racial Reckoning" by Tamara Lee, Maite Tapia, Naomi Williams, and Sheri Davis Faulkner. We have a 2023 Research Volume being discussed and will be brought to you at the January 2022 board meeting for review. The LERA/ILR Special Issue process from the 2020 annual meeting resulted in nine paper submissions. Unfortunately, none of the papers made it all the way through this process this year, but the editorial committee recommends continuing this process. This year's call will have a deadline of June 30, 2021.

Program Committee Report—Paul Clark reports that the program committee met and came away excited about the prospect of meeting in Detroit, Michigan in 2022. The theme focuses on "Elevating Voice and New Voices in the Workplaces and Beyond", referencing the desire of workers to have input into their work, and recognizing that not all have had this opportunity equally. John Budd has been asked to be the academic vice chair and Bill Spriggs has been invited to be the practitioner vice chair. Our regional vice chair will be Robert Chiaravalli, in residence in the Detroit area and long connected to the Detroit LERA Chapter. We talked in the committee about all the usual components of the program, and the year ahead certainly promises to be an interesting year in the field of labor and employment relations, with no shortage of issues to focus on. The committee will do our best to aggressively encourage people to put together proposals and build a program. In terms of actually being in Detroit, Robert Chiaravalli had some great ideas about special events, focusing on the labor history of the automobile industry. Site visits are a possibility. Wayne State houses a labor history archives. A MoTown tour was mentioned. Lastly, it was pointed out that we are just across the bridge from neighboring Canada, and we hope by next June the border will be open, and hopefully that being in such proximity to Canada may encourage participation from Canadians. Wilma Liebman mentioned that she interviewed the Secretary of Labor and invited him to come back and visit us again in person in Detroit, and report on his first year's progress. David Lewin asks if there will be sessions on voice at Google and/or Amazon as they both have on-going organization efforts.

Teaching Interest Section Report—Paul Clark also reported progress on creating a teaching interest section, a group that can meet periodically and share case studies. Adrienne Eaton asked to have a session put together on the meeting program next week, which has been done, and an organizing committee has been established. An organizational meeting will take place on Sunday, June 6, 2021 at 12:15 p.m. There will also be a session at 1:45 p.m. on innovative approaches to online teaching in employment relations as well. This will be in a roundtable format where people can share their ideas.

LERA@ASSA Program—The program committee, led by William Rodgers at Rutgers University, for the LERA@ASSA has put together 18 sessions again for the January 2022 meeting which will take place in Boston. We should really emphasize that the LERA sessions at the ASSA meeting are quite accessible and widely appealing to sociologists and others. That is a meeting worth attending for a broader audience than just labor economists. The idea was circulated to perhaps move the board to one virtual meeting and one in-person meeting each year, to coincide with the LERA Annual Meeting. There was no adverse opinion to moving the January board meeting to a virtual format.

Development Committee Report—Harry Katz reports that development funds have contributed to the solid financial state of LERA reviewed by our secretary-treasurer. We raised $36,000 last year in individual contributions, and $40,000 from our major sustaining sponsors, Ford Motor Co., and Kaiser Permanente. We are also building a quasi-endowment fund focusing on major gifts and final gifting, so please think about this in your estate planning. One thing you can help us on, we are always looking for additional sustaining sponsors, and if you have tips on that, Harry Katz and Jim Pruitt are happy to make the ask. Jim Pruitt made a personal request that each individual consider donating a $1,000 each year, depending upon individual personal ability to do so.

Industry Councils-Interest Sections Coordinating Committee Report—Joel Cutcher-Gershenfeld, Chair, reported that we have eight established interest sections and one forming, the Teaching Interest Section. We also have eight established industry councils and three new ones being proposed. Many of these groups are represented on the annual meeting program. When LERA members join or renew their membership they indicate which interest sections they want to affiliate with, and which industry councils, if any, with which they want to affiliate and he reported on current numbers in all these LERA groups. These numbers give a pulse of the membership and the numbers are substantial. Even 30-50 people represent solid numbers, let alone 100-1000 people that have expressed interest in a specific topic. We will meet at one o'clock on Saturday with all the co-chairs of the interest sections and industry councils. Last year, we had a number of webinars organized by industry councils and interest sections that brought in many new LERA members, and we will be considering what mix of programming and services should be offered in the future. Our website allows our industry council and interest section communities to freely and autonomously communicate with each other, which should help these groups be more active going forward. We thank Paul Clark for taking a leadership role in the new Teaching Interest Section, which held its first meeting this year in conjunction with the LERA 73rd Annual Meeting.

Nominating Committee Report— Emily Smith reported on behalf of Rose Batt, Committee Chair, who could not be with us today. The slate of candidates being reviewed by the board today represent two primary nominees for each seat, and a list of four alternates in each category, in the event that someone in the primary nominee slot would decline the invitation to run for the board. Experience has taught us that some will not wish to run against a close colleague, or it might not be the right time professionally to run, so we have a list of alternates. The slate is confidential, and all the names on this slate represent a qualified and diverse group of leaders who all have the potential to lead this organization. The candidates will be invited to run following the annual meeting, and we expect to conduct a 2021 summer election. The winners will assume office following the June meeting in 2022, and the next president elect who will run unopposed on the ballot will be Bill Spriggs of the AFL-CIO. With one change to the slate in the category of Regional Vice President, the slate passed unanimously by the board. The change in the RVP category is as follows: Kim LaFevor will be invited to run for election against Robert Chiaravalli, who currently occupies this seat.

Membership Committee Report—David Lewin reports that the January meeting that continues with the ASSA. When ASSA decided to stop paying a portion of registration dues to us, we decided to move our annual meeting to our own format in January. We now have more meeting sessions if we combined January and June sessions which allows greater opportunities for members and others to participate; this is a great membership benefit. Our membership since 2015 has been increasing, though we are down in 2021 by about 30 members. Suggestions from this morning's membership committee meeting would be to reach out to the UCIRHRP to try to attract labor relations programs from academic institutions, and another area is to reach out to local chapters to become members of national LERA. The committee feels that member dues are appropriate and should not be increased for another year. Adrienne Eaton discussed progress on the task force that was created at the last board meeting. Erin Johannson was focused on recruiting labor members and those from non-traditional labor movements. The other was a task force to recruit management members, led by Kevin Legel and Bill Dwyer, and they have a letter that can be used to recruit of management members. There is more work to be done, and we can report back at the next board meeting. Kevin Legel suggested sharing this template letter with the entire board. Jim Pruitt will work on UHW to interest them in membership in LERA. Joel Cutcher-Gershenfeld suggests that there used to be many labor-management entities, and LERA may be the last labor-management entity standing. There are many polarized organizations that are important and do great work, but LERA is the last place to build relationships across all parties and perspectives. Bill Canak asks if Ford Motor Co. and Kaiser Permanente might be an appropriate place to reach out to HR and labor leaders to introduce the idea of LERA being their professional home. There needs to be a value proposition in place for these groups that is clear and worthwhile for these groups. Tamara Lee asks about a membership fee structure or partnerships between ALRA or LAWCHA or other groups that have an affinity with our purposes. Tamara Lee volunteers to work on linkages between these groups.

Nominating Committee Report—Emily Smith reports on behalf Rose Batt, Committee Chair, who could not be with us today. The slate being reviewed by the board today represents primary nominees, and a selection of alternates in each category since experience tells us that some individuals may decline our invitation to run for the executive board as they may not wish to run against a close colleague. The slate put together by the committee is diverse, inclusive, and all the people are qualified to lead the organization. If elected, terms for this slate would begin in 2022 following the June meeting, and the next president, who would run unopposed, would be Bill Spriggs of the AFL-CIO if the board approves. With the addition of one change, the board passed the slate unanimously, motioned by Bill Canak. The change affects the seat of the regional vice president; the two nominees who will be invited to run are Kim LaFevor and Robert Chiaravalli, who currently occupies this seat.

National Chapter Advisory Council Report—William Canak, NCAC Chair, reports that the state of LERA chapters is good despite the pandemic. We have a considerable number of chapters receiving LERA star awards, outstanding chapter awards, and merit awards, which is a reflection of their engagement. A small number are a little tardy on their annual dues, and we expect those to come in shortly. A new chapter has been formulated in Virginia, which is really coming along. The monthly LERA eBulletin has done a good job reporting on chapter activity and events. We are looking forward to another VP election this summer, and Kim LaFevor and Robert Chiaravalli will be running for the Mid Regional Vice President seat. The NCAC meeting this morning was well-attended, and the upcoming LERA 73rd Annual Meeting will feature several chapter specific sessions and events. The bylaw changes in 2016 enacted by LERA have changed the whole nature of relationships between LERA and our chapters. Innovations in chapter programs and membership will be discussed at the representatives meeting, and NorCal LERA and NJ LERA will focus on the work they have done over the last few years. Chapters look forward to having more LERA members join their local chapters as is required by LERA bylaws. Bonnie Castrey asks the regional vice presidents to be in close contact with the chapters in their regions, and to remind the chapters in their region to share their email lists with national so that their chapter members receive affiliate benefits.

Regional Vice Presidents Report—Robert Chiaravalli has done a review of the chapters in his area. Membership ranges from 50-100 members per chapters, conducting in person and virtual meetings, and treasuries are healthy. The biggest challenges are participation and membership especially in light of the pandemic. All chapters are doing things a little differently, and some of these innovations are showing progress. Beverly Harrison reports that her chapters are financially stable. Challenges they are experiencing include technical glitches with Zoom, not collecting dues this year, and missing the camaraderie and networking of in-person event. Long Island has 500 members and a $17,000 bank balance. The newest chapter forming is Virginia LERA. She has already formulated a diverse array of officers, and has held two virtual meetings since the Fall. The Buffalo chapter was revived in 2019. Jim Pruitt reports that NorCal LERA has been re-invigorated. Oregon LERA has dealt well with the virtual world we all entered into last year. Burton White passed last year from the Oregon LERA family.

Diversity and Inclusion Report—Virginia Doellgast, chair, reports that the committee met earlier today and the recommendations made at the last meeting have been implemented. The first recommendation was to give LERA members the opportunity to report on their gender and race, and we are starting to get more data so we can begin tracking to ultimately increase diversity. We now have information in the CFP that the committee will look at diversity when deciding panels, and the program committee will be focusing on diverse plenaries and featured speakers. Next year, the LERA submission form will ask panel organizers to comment on the diversity of their panel, and an ability to indicate if some of their session participants may require financial assistance. This will give the program committee a better idea about which sessions are diverse, and which may require additional financial assistance. LERA is informing awardees and volunteer leaders of our requirements about sexual harassment. Next year, we will try to not leave this item to the end of the agenda. Last year, the Gershenfeld fund was re-purposed to support diversity and inclusion at LERA annual meetings. Additionally, the board has set aside registration funds to additionally support registration at the annual meeting for those who have a financial need and increase diversity and inclusion at the annual meeting.

Next LERA executive board meeting will take place virtually in February 2022.

Adjournment—The meeting was adjourned at 7:41 p.m. Eastern Time by President Adrienne Eaton.

LERA General Membership Meeting Minutes Sunday, June 6, 2021, 3 p.m. Eastern Time Via Videoconference

Call to order—Adrienne Eaton opened at 3 p.m. Eastern Time. She thanked Wilma Leibman, Program Committee Chair and President Elect. She also thanked the Program Committee Vice Chairs, Lynne Rhinehart, Aaron Sojourner, and Robert Chiaravalli, and the LERA staff. Since we have a limited amount of time for our meeting, we will keep our reports short.

Financial Report—Andrew Weaver, Secretary-Treasurer, reported that 2020 was a challenging year, but LERA performed well despite the circumstances. There were revenue reductions due to the virtual conference, but expenses were reduced accordingly so the organization is still in the black with about a six percent net income. Given what has happened to most small non-profits over the last year, we think this is solid performance. Our balance statement is good with a decent amount of cushion.

Membership Committee Report—David Lewin, Chair, reported that our membership has been increasing since 2015, but with a slight decrease this year, which we believe is due to an impact of the pandemic. The membership has been solid in the last several years, and there are a number of initiatives underway in order to enhance the membership numbers going forward.

Development Committee Report—Jim Pruitt, Co-Chair, reported that 2020 was a strong fund-raising year. We received both restricted and non-restricted contributions, and 2020 was stronger than 2019 and other previous years. LERA would like people to consider a personal gift to LERA. If you are in a situation where you can afford a contribution, I would ask you to consider a $1,000 gift to LERA, or more or less as appropriate. There are those who are listening who have already contributed more than this, and we are truly grateful. This is a wonderful organization, it provides for the neutral and advocates on either side to be together in a spirit of collegiality and teamwork. Also, the staff is efficient, and the organization spends money well. Jim thanked the organizations who are sustaining sponsors, Kaiser Permanente and Ford Motor Co., and encouraged other organizations to follow suit. LERA also has many institutions of higher education that provide a backbone of support to the LERA meeting and organization in various ways.

Editorial Committee Report—Ryan Lamare, Editor-in-Chief, thanked the editorial committee that met last week. The 2021 LERA Research Volume, edited by Tobias Schulze-Cleven and Todd Vachon on "Revaluing Work(ers): Toward Democratic and Sustainable Prosperity" is about to be released. The 2022 Volume is "Industrial Relations and a Racial Reckoning" edited by Maite Tapia, Tamara Lee, Naomi Williams, and Sheri Davis is in production for release in 2022. Those of you presenting a paper at the LERA 73rd Annual Meeting this year should consider participating in the LERA/ILR Review Best Paper Competition, and the deadline is June 30, 2021 to complete your submission, which is available at the LERA website.

Program Committee Report—Paul Clark, Program Committee Chair and President Elect Elect, reported that the program committee selected is a talented group. The theme of the program is "Elevating Voice and New Voices in the Workplace and Beyond". John Budd and Bill Spriggs will be the Vice Chairs. Robert Chiaravalli, from Detroit LERA, will lead a regional team who will be working on site visits and special events. As you can imagine there are a number of sites and historical places of relevance in Detroit, and we are excited to bring this event to life over the next year. We hope all of you listening will be thinking about proposals to submit by November 15, 2021 for the program committee to review. We are definitely planning to be back in person in Detroit 2022, but we are also want to try to keep at least a few of the elements from this virtual way of meeting.

Diversity and Inclusion Committee Report—Emily Smith, LERA Executive Director, reported on behalf of Virginia Doellgast, Chair. LERA is making strides towards reporting on gender and race, and so we can begin tracking to ultimately increase diversity. We now have information in the CFP that the committee will look at diversity when deciding panels, and the program committee will be focusing on diverse plenaries and featured speakers. Next year, the LERA submission form will ask panel organizers to comment on the diversity of their panel, and an ability to indicate if some of their session participants may require financial assistance. This will give the program committee a better idea about which sessions are diverse, and which may require additional financial assistance. LERA is informing awardees and volunteer leaders of our requirements about sexual harassment findings as well. The Gershenfeld fund has been re-purposed to support diversity and inclusion at LERA annual meetings and giving to this fund last year increased when this was announced.

National Chapter Advisory Council Report—by Bill Canak, Chair and Bonnie Castrey, Vice Chair. The state of the chapters is healthy; three chapters are receiving star awards, and many are receiving outstanding chapter or merit awards. The LERA eBulletin has been reporting on chapter events, which has taken considerable work on behalf of the staff and has LERA a more cohesive organization. Our Regional Vice Presidents have made a good impact, and you will see one one seat on this summer's election ballot. Tomorrow we will have a chapter representative meeting which has comprehensive chapter participation around the country, and there will also be a chapter administration workshop. If you are a national member of LERA and not currently a chapter member, then please get in touch with NCAC leaders and we will help you get connected with a local chapter in your area.

Industry Councils and Interest Sections Coordinating Committee Report—Joel Cutcher-Gershenfeld, Chair, reported that we have eight established interest sections and one forming, the Teaching Interest Section. We also have eight established industry councils and three new ones being proposed. Many of these groups are represented on the annual meeting program. Last year, we had a number of webinars organized by industry councils and interest sections that brought in many new LERA members, and we will be considering what mix of programming and services should be offered in the future. Our website allows our industry council and interest section communities to freely and autonomously communicate with each other, which should help these groups be more active going forward. We thank Paul Clark for taking a leadership role in the new Teaching Interest Section, which held its first meeting this year in conjunction with the LERA 73rd Annual Meeting.

LERA Awards Committee Report—Paul Clark, Co-Chair, reported on the LERA 2021 Awards Winners. The capstone Lifetime Achievement Award was awarded to Dennis Dabney, Kris Rondeau, and Paula Voos. The Academic Fellows were Annette Bernhardt, Alex Colvin, and Jack Fiorito. Practitioner Fellows were Richard Fincher, Jennifer Kelly, and Fred Stahl. The Susan C. Eaton Outstanding Scholar-Practitioner Award was conferred on C. Jeffrey Waddoups. The Myron C. Taylor Management Award was awarded to Edward Potter. The John T. Dunlop Outstanding Scholar Awards were given to Emily Twarog for contributions of national significance, and to Christian Ibsen for contributions of international significance. The LERA Outstanding Practitioner Award was awarded to Lisa Charles. The James G. Scoville Best International/Comparative IR/HR Paper Award was received by Sarah Ashwin. The Kenneth May Media Award was bestowed on Hamilton Nolan, Jeff Schuhrke, Lauren Kaori Gurley, and Noam Scheiber. This year's Thomas A. Kochan and Stephen R. Sleigh Best Dissertation Award went to two separate dissertation authors: Yao Yao and Xiaoming Bao. Th UCIRHRP Best Student Paper award winner was Hyesook Chung. There is an awards session tomorrow at 11:15 am Eastern Time tomorrow where we will hear from several of the 2021 Award Winners where we will both recognize them and listen to them.

President Adrienne Eaton concluded the meeting with a few announcements, including that the LERA 74th Annual Meeting will take place in Detroit, Michigan June 2-5, 2022, and we will also meet in Boston, MA January 7-9, 2022 with the ASSA (LERA@ASSA 2022 Meeting). However, the board will meet virtually instead of in person at Boston.

Incoming President Wilma Liebman thanked standing President Adrienne Eaton for her leadership over the last year, especially under difficult circumstances, and Adrienne Eaton received a clock commemorating her leadership of this organization. Wilma said Adrienne Eaton is the essence of what the organization is about and the first virtual LERA president, and hopefully the last, and Wilma thanked her for getting us through this difficult period. Wilma praised his program committee and vice chairs, and Steve Greenhouse, for going above and beyond on session organizing. Wilma thanked all of them, and hopes everyone is enjoying the meeting. Wilma also invites all past LERA presidents to contact her with any input, adjustments, criticisms, or other recommendations as she enters into her presidential year. Wilma recognizes what a unique organization this is, bringing together so many constituencies. Wilma will have more to say about Adrienne tomorrow at the presidential address, but she had one comment to make today. "Adrienne's career and contributions are the essence of what LERA is all about. In addition to being a first-rate scholar, she is president of a faculty union, and the Dean of her department. She also serves the public in various ways, and is a "triple-threat academic", a Scholar, a Practitioner, and a Teacher." Wilma looks forward to working with everyone in the year to come.

Adjournment—The meeting was adjourned at 3:56 p.m. Eastern Time.

Labor and Employment Relations Association

Champaign, Illinois

Financial Statements

For the Years Ended December 31, 2020 and 2019

Feller & Kuester CPAs LLP
Certified Public Accountants
806 Parkland Court
Champaign, IL 61821
217-351-3192

Feller & Kuester CPAs LLP

Tax - Audit - Bookkeeping
806 Parkland Court - Champaign, Illinois 61821
Phone - (217) 351-3192 Fax - (217) 351-4135 Email - neal@fellerkuester.com

INDEPENDENT AUDITOR'S REPORT

To the Board of Directors of
Labor and Employment Relations Association
Champaign, Illinois

We have audited the accompanying financial statements of the Labor and Employment Relations Association (a nonprofit organization) which comprise the statements of financial position as of December 31, 2020 and 2019, and the related statements of activities, functional expenses, and cash flows for the years then ended, and the related notes to the financial statements.

Management's Responsibility for the Financial Statements
Management is responsible for the preparation and fair presentation of these financial statements in accordance with accounting principles generally accepted in the United States of America; this includes the design, implementation, and maintenance of internal control relevant to the preparation and fair presentation of financial statements that are free from material misstatement, whether due to fraud or error.

Auditor's Responsibility
Our responsibility is to express an opinion on these financial statements based on our audits. We conducted our audits in accordance with auditing standards generally accepted in the United States of America. Those standards require that we plan and perform the audit to obtain reasonable assurance about whether the financial statements are free from material misstatement.

An audit involves performing procedures to obtain audit evidence about the amounts and disclosures in the financial statements. The procedures selected depend on the auditor's judgment, including the assessment of the risks of material misstatement of the financial statements, whether due to fraud or error. In making those risk assessments, the auditor considers internal control relevant to the entity's preparation and fair presentation of the financial statements in order to design audit procedures that are appropriate in the circumstances, but not for the purposes of expressing an opinion on the effectiveness of the entity's internal control. Accordingly, we express no such opinion. An audit also includes evaluating the appropriateness of accounting policies used and the reasonableness of significant accounting estimates made by management, as well as evaluating the overall presentation of the financial statements.

We believe that the audit evidence we have obtained is sufficient and appropriate to provide a basis for our audit opinion.

Opinion

In our opinion, the financial statements referred to above present fairly, in all material respects, the financial position of the Labor and Employment Relations Association as of December 31, 2020 and 2019, and the changes in its net assets and its cash flows for the years then ended in accordance with the accounting principles generally accepted in the United States of America.

Feller & Kuester CPAs LLP

Feller & Kuester CPAs LLP
Champaign, Illinois

May 28, 2021

LABOR AND EMPLOYMENT RELATIONS ASSOCIATION
STATEMENTS OF FINANCIAL POSITION
DECEMBER 31, 2020 AND 2019

	2020	2019
ASSETS		
Current Assets		
Cash and Cash Equivalents	$ 325,553	$ 283,130
Investments	91,321	67,733
Accounts Receivable, Net	-	-
Prepaid Expenses	8,617	17,600
Inventory	6,685	7,264
Accrued Royalties	-	-
Total Current Assets	432,176	375,727
Property and Equipment, Net	2,728	4,367
TOTAL ASSETS	$ 434,904	$ 380,094
LIABILITIES AND NET ASSETS		
Current Liabilities		
Accounts Payable	$ 750	$ -
Accrued Liabilities	42,055	34,322
Dues Collected in Advance	72,986	70,789
Subscriptions Collected in Advance	11,418	12,066
Deferred Chapter Dues	7,500	5,600
Other Deferred Revenue	23,175	37,175
Total Current Liabilities	157,884	159,952
Net Assets		
Without Donor Restrictions		
Designated by the Board		
Susan C. Eaton Scholar-Practitioner Memorial Fund	20,779	23,779
Kochan-Sleigh Best Dissertation Award Fund	37,439	38,676
Jim Scoville Award Fund	3,296	3,924
John T. Dunlop Public Policy Fund	19,005	16,840
Gladys and Water Gershenfeld Publication Fund	17,147	17,300
Quasi-Endowment Fund	49,190	32,448
Total Designated by the Board	146,856	132,967
Undesignated	107,133	73,664
Total Without Donor Restrictions	253,989	206,631
With Donor Restrictions	23,031	13,511
Total Net Assets	277,020	220,142
TOTAL LIABILITIES AND NET ASSETS	$ 434,904	$ 380,094

See Accompanying Notes.

LABOR AND EMPLOYMENT RELATIONS ASSOCIATION
STATEMENT OF ACTIVITIES
FOR THE YEAR ENDED DECEMBER 31, 2020

	Without Donor Restrictions			With Donor	
	Undesignated	**Designated**	**Total**	**Restrictions**	**Total**
Support and Revenues					
Membership Dues	$ 149,333	$ -	$ 149,333	$ -	$ 149,333
Meeting Income	109,655	-	109,655	9,060	118,715
Organization Dues & Sponsorships	62,196	-	62,196	460	62,656
Subscriptions	16,031	-	16,031	-	16,031
Chapter Fees	10,328	-	10,328	-	10,328
Member Contributions	44,672	33,584	78,256	-	78,256
Royalties	6,549	-	6,549	-	6,549
Ad Income	9,880	-	9,880	-	9,880
Publications	100	-	100	-	100
Administrative Fees	520	-	520	-	520
Investment Income	(4,211)	(14,181)	(18,392)	-	(18,392)
Interest Income	460	-	460	-	460
Net Assets Released from Designation	5,514	(5,514)	-	-	-
Total Support and Revenues	411,027	13,889	424,916	9,520	434,436
Expenses					
Program Services					
General	219,005	-	219,005	-	219,005
Meetings	31,796	-	31,796	-	31,796
Publications	41,476	-	41,476	-	41,476
Supporting Services					
Management and General	73,798	-	73,798	-	73,798
Membership Development	11,483	-	11,483	-	11,483
Total Expenses	377,558	-	377,558	-	377,558
Change in Net Assets	33,469	13,889	47,358	9,520	56,878
Net Assets, Beginning of Year	73,664	132,967	206,631	13,511	220,142
Net Assets, End of Year	$ 107,133	$ 146,856	$ 253,989	$ 23,031	$ 277,020

See Accompanying Notes.

LABOR AND EMPLOYMENT RELATIONS ASSOCIATION
STATEMENT OF ACTIVITIES
FOR THE YEAR ENDED DECEMBER 31, 2019

	Without Donor Restrictions			With Donor	
	Undesignated	Designated	Total	Restrictions	Total
Support and Revenues					
Membership Dues	$ 151,409	$ -	$ 151,409	$ -	$ 151,409
Meeting Income	194,298	-	194,298	206	194,504
Organization Dues & Sponsorships	62,492	-	62,492	280	62,772
Subscriptions	17,848	-	17,848	-	17,848
Chapter Fees	9,636	-	9,636	-	9,636
Member Contributions	35,841	31,360	67,201	-	67,201
Royalties	5,236	-	5,236	-	5,236
Ad Income	9,050	-	9,050	-	9,050
Publications	1,282	-	1,282	-	1,282
Administrative Fees	70	-	70	-	70
Investment Income	272	3,598	3,870	-	3,870
Interest Income	865	-	865	-	865
Net Assets Released from Designation	4,880	(4,880)	-	-	-
Total Support and Revenues	493,179	30,078	523,257	486	523,743
Expenses					
Program Services					
General	198,684	-	198,684	-	198,684
Meetings	118,840	-	118,840	-	118,840
Publications	39,340	-	39,340	-	39,340
Supporting Services					
Management and General	70,353	-	70,353	-	70,353
Membership Development	19,076	-	19,076	-	19,076
Total Expenses	446,293	-	446,293	-	446,293
Change in Net Assets	46,886	30,078	76,964	486	77,450
Net Assets, Beginning of Year	26,778	102,889	129,667	13,025	142,692
Net Assets, End of Year	$ 73,664	$ 132,967	$ 206,631	$ 13,511	$ 220,142

See Accompanying Notes.

LABOR AND EMPLOYMENT RELATIONS ASSOCIATION
STATEMENT OF FUNCTIONAL EXPENSES
FOR THE YEAR ENDED DECEMBER 31, 2020

		MEETINGS			PUBLICATIONS			SUPPORTING SERVICES		
	General	Annual Conference	ASSA Meeting	Other Meetings	Perspectives	Research Volume	Other Publications	Management & General	Membership Development	Total
Compensation	$ 143,096	$ -	$ -	$ -	$ -	$ -	$ -	$ 28,191	$ -	$ 171,287
Payroll Taxes & Fringes	57,459	-	-	-	-	-	-	10,681	-	68,140
Contract Labor	18,450	-	-	-	-	-	-	-	-	18,450
Depreciation	-	-	-	-	-	-	-	1,639	-	1,639
Insurance	-	-	-	-	-	-	-	892	-	892
Bank and Service Charges/Fulfillment	-	-	-	-	-	-	-	16,127	-	16,127
Promotion	-	-	-	-	-	-	-	-	-	-
Postage/Freight	-	-	90	-	2,124	2,600	-	387	-	5,201
Accounting/Auditing	-	-	-	-	-	-	-	6,707	-	6,707
Printing and Production	-	-	171	-	8,460	7,305	18	-	-	15,954
Services	-	14,000	-	-	11,738	8,605	552	-	-	34,895
Other Publication Costs	-	-	-	-	74	-	-	-	-	74
Meals/Receptions	-	2,000	10,110	-	-	-	-	-	-	12,110
Travel	-	1,267	2,239	539	-	-	-	-	-	4,045
Other Meeting Expenses	-	1,065	315	-	-	-	-	-	-	1,380
Computer Supplies/Services	-	-	-	-	-	-	-	896	-	896
Website	-	-	-	-	-	-	-	4,410	-	4,410
Office Supplies	-	-	-	-	-	-	-	3,758	-	3,758
Student & Member Awards	-	-	-	-	-	-	-	-	7,107	7,107
Fundraising Expense	-	-	-	-	-	-	-	-	1,602	1,602
Duplicating Expense	-	-	-	-	-	-	-	-	-	-
Chapter Expenses	-	-	-	-	-	-	-	-	1,774	1,774
Other Committee Expenses	-	-	-	-	-	-	-	-	1,000	1,000
Miscellaneous Office	-	-	-	-	-	-	-	110	-	110
Total	$ 219,005	$ 18,332	$ 12,925	$ 539	$ 22,396	$ 18,510	$ 570	$ 73,798	$ 11,483	$ 377,558

See Accompanying Notes.

LABOR AND EMPLOYMENT RELATIONS ASSOCIATION
STATEMENT OF FUNCTIONAL EXPENSES
FOR THE YEAR ENDED DECEMBER 31, 2019

		MEETINGS			PUBLICATIONS			SUPPORTING SERVICES		
	General	Annual Conference	ASSA Meeting	Other Meetings	Perspectives	Research Volume	Other Publications	Management & General	Membership Development	Total
Compensation	$ 129,378	$ -	$ -	$ -	$ -	$ -	$ -	$ 27,586	$ -	$ 156,964
Payroll Taxes & Fringes	51,654	-	-	-	-	-	-	9,712	-	61,366
Contract Labor	17,652	-	-	-	-	-	-	-	-	17,652
Depreciation	-	-	-	-	-	-	-	539	-	539
Insurance	-	-	-	-	-	-	-	923	-	923
Bank and Service Charges/Fulfillment	-	-	-	-	-	-	-	11,139	-	11,139
Promotion	-	6,569	120	-	-	-	-	-	1,472	8,161
Postage/Freight	-	-	68	6	2,219	2,176	-	1,079	-	5,548
Accounting/Auditing	-	-	-	-	-	-	-	5,015	-	5,015
Printing and Production	-	3,389	-	-	8,324	7,879	-	-	-	19,592
Services	-	2,940	-	-	7,679	9,515	1,176	-	-	21,310
Other Publication Costs	-	-	-	-	230	109	33	-	-	372
Meals/Receptions	-	57,501	5,901	50	-	-	-	-	-	63,452
Travel	-	5,605	2,345	580	-	-	-	-	-	8,530
Other Meeting Expenses	-	32,458	1,308	-	-	-	-	-	-	33,766
Computer Supplies/Services	-	-	-	-	-	-	-	10,265	-	10,265
Website	-	-	-	-	-	-	-	2,030	-	2,030
Office Supplies	-	-	-	-	-	-	-	2,059	-	2,059
Student & Member Awards	-	-	-	-	-	-	-	-	7,056	7,056
Fundraising Expense	-	-	-	-	-	-	-	-	5,933	5,933
Duplicating Expense	-	-	-	-	-	-	-	6	-	6
Chapter Expenses	-	-	-	-	-	-	-	-	4,594	4,594
Other Committee Expenses	-	-	-	-	-	-	-	-	21	21
Miscellaneous Office	-	-	-	-	-	-	-	-	-	-
Total	$ 198,684	$ 108,462	$ 9,742	$ 636	$ 18,452	$ 19,679	$ 1,209	$ 70,353	$ 19,076	$ 446,293

See Accompanying Notes.

LABOR AND EMPLOYMENT RELATIONS ASSOCIATION
STATEMENTS OF CASH FLOWS
FOR THE YEARS ENDED DECEMBER 31, 2020 AND 2019

	2020	2019
CASH FLOWS FROM OPERATING ACTIVITIES		
Change in Net Assets	$ 56,878	$ 77,450
Adjustments to Reconcile Change in Net Assets to Net Cash Provided by (Used in) Operating Activities:		
Depreciation Expense	1,639	539
Net Unrealized (Gains) Losses on Investments	22,303	(3,140)
(Increase) Decrease in Operating Assets:		
Accounts Receivable, Net	-	583
Prepaid Expenses	8,983	(9,384)
Inventory	579	(890)
Accrued Royalties	-	2,579
Increase (Decrease) in Operating Liabilities:		
Accounts Payable	750	(82)
Accrued Liabilities	7,733	3,727
Dues Collected in Advance	2,197	(1,965)
Subscriptions Collected in Advance	(648)	(782)
Deferred Chapter Dues	1,900	550
Other Deferred Revenue	(14,000)	23,475
Net Cash Provided by (Used in) Operating Activities	88,314	92,660
CASH FLOWS FROM INVESTMENT ACTIVITIES		
Donated Investments	(41,980)	(39,200)
Reinvested Interest, Dividends, and Capital Gains	(3,911)	(730)
Purchases of Property and Equipment	-	(3,398)
Net Cash Provided by (Used in) Investment Activities	(45,891)	(43,328)
CASH FLOWS FROM FINANCING ACTIVITIES		
None	-	-
Net Cash Provided by (Used in) Financing Activities	-	-
Net Increase (Decrease) in Cash and Cash Equivalents	42,423	49,332
Beginning Cash and Cash Equivalents	283,130	233,798
Ending Cash and Cash Equivalents	$ 325,553	$ 283,130

See Accompanying Notes.

NOTE 1 – NATURE OF ACTIVITIES AND SIGNIFICANT ACCOUNTING POLICIES

Organization and Nature of Activities

The Labor and Employment Relations Association (the Organization) was founded in 1947 to encourage research in all aspects of the field of labor, employment, and the workplace. It is a nonprofit scholarly association of academic, labor, business, and neutral communities committed to the full discussion and exchange of ideas between and among its broad constituencies through meetings, publications, and its various electronic listservs and websites. The Organization's National Office is located in Champaign, Illinois and serves the association by planning conferences and meetings and publishing the various researches of its members. The main sources of support and revenues for the Organization are contributions, membership dues and subscriptions, and meeting income.

Subsequent Events

Subsequent events have been evaluated through May 28, 2021, which is the date the financial statements were available to be issued.

Accrual Basis of Reporting

The Organization has chosen to report on the accrual basis of accounting. Accordingly, revenue is recognized when earned and expenses are recognized when incurred in conformity with accounting principles generally accepted in the United States of America (GAAP). The financial statements are presented in accordance with Financial Accounting Standards Board (FASB) Accounting Standards Codification (ASC) 958 dated August 2016, and the provisions of the American Institute of Certified Public Accountants (AICPA) "Audit and Accounting Guide for Not-for-Profit Organizations" (the "Guide").

Estimates

The preparation of financial statements in conformity with GAAP requires management to make estimates and assumptions that affect certain reported amounts and disclosures. Accordingly, actual results could differ from those estimates.

Cash and Cash Equivalents

For purposes of reporting cash flows, cash and cash equivalents include all cash and highly liquid investments acquired with an original maturity date of three months or less. Since the penalties of converting certificate of deposits to cash is insignificant, all certificate of deposits have been included with cash and cash equivalents. As of December 31, 2020 and 2019, $67,832 and $73,346 of cash and cash equivalents are designated by the board and are subject to board-imposed stipulations. As of December 31, 2020 and 2019, $23,031 and $13,511 of cash and cash equivalents are temporarily restricted and are subject to donor stipulations.

Investments

Investments consist of intermediate term bond funds, Bayer AG (BAYRY) stock, and money market funds. Net appreciation (depreciation) in the fair value of investments, which consists of the realized gains or losses and the unrealized appreciation (depreciation) on those investments, is presented in the statement of activities in accordance with donor restrictions as investment income. Investment income is presented net of investment fees. The average cost method is primarily used to determine the basis for computing realized gains or losses.

Accounts Receivable

Accounts receivable are recorded primarily for outstanding invoices for membership dues and subscriptions. An allowance for doubtful accounts is based on an analysis of expected collection rates determined from experience. The Organization had an allowance for doubtful accounts of $0 and $0 at December 31, 2020 and 2019, respectively.

Inventory

The Organization's inventory of directories, research volumes, proceedings, and perspective magazines is carried at the lower of cost and market value. Cost is determined on the basis of first in – first out.

Property and Equipment

Property and equipment expenditures in excess of $1,000 are capitalized at cost. Donated property and equipment are capitalized at estimated cost or fair market value at the time of donation. Depreciation of the assets is computed using the straight-line method over their estimated useful lives. The range of estimated useful lives by type of asset is as follows:

Furniture and Equipment	3 years

Net Assets

Net assets of the Organization and changes therein are classified and reported as follows:

Net Assets without Donor Restrictions - Net assets that are not subject to donor-imposed restrictions and may be expended for any purpose in performing the primary objectives of the Organization. The Organization's board may designate assets without restrictions for specific operational purposes from time to time.

Net Assets with Donor Restrictions - Net assets subject to stipulations imposed by donors, and grantors. Some donor restrictions are temporary in nature; those restrictions will be met by actions of the Organization or by the passage of time. Other donor restrictions are perpetual in nature, where by the donor has stipulated the funds be maintained in perpetuity. As of December 31, 2020 and 2019, the Organization had no donor restrictions that were perpetual in nature.

Contributions

Unconditional contributions are recognized when pledged and recorded as net assets without donor restrictions or net assets with donor restrictions, depending on the existence and/or nature of any donor-imposed restrictions. Conditional promises to give are recognized when the conditions on which they depend are substantially met. Gifts of cash and other assets are reported with donor restricted support if they are received with donor stipulations that limit the use of the donated assets.

When a restriction expires, that is, when a stipulated time restriction ends or a purpose restriction is accomplished, net assets with donor restrictions are reclassified to net assets without donor restrictions and reported in the statement of activities as net assets released from restrictions. Donor-restricted contributions whose restrictions are met in the same reporting period are reported as net assets without donor restriction support. Contributions restricted for the acquisition of land, buildings, and equipment are reported as net assets without donor restriction upon acquisition of the assets and the assets are placed in service.

Contributed Services and Goods

Contributed services are reported as contribution revenue and as assets or expenses only if the services create or enhance a non-financial asset (for example property and equipment) or:

- Would typically need to be purchased by the Organization if the services had not been provided by contribution.
- Require specialized skills.
- Are provided by individuals with those skills (such as accounting, financial, construction, educational, electrical, legal, medical, and other services provided by accountants, investments advisers, contractors, teachers, electricians, lawyers, doctors, and other professional and craftspeople).

For the years ended December 31, 2020 and 2019, the value of contributed services meeting the defined requirements for recognition in the financial statements as outlined above were not material and have not been recorded on the financial statements. The Organization does receive free office space from the University of Illinois. However, the value for the use of this office space has not been reported on the financial statements.

Income Tax Status

The Organization is a nonprofit association that is exempt from federal income tax under Section 501(c)(3) of the Internal Revenue Code. In addition, the Internal Revenue Service has determined that the Organization is not a private foundation as defined in Section 509(a)(1) and Section 170(b)(1)(A)(vi) of the Code.

The Organization has evaluated its exposure resulting from uncertain income tax position and determined the exposure is not material to the financial statements. In addition, the Organization is not aware of any tax position for which a significant change is reasonably possible within the next 12 months. Therefore, these financial statements do not include a liability for uncertain tax positions. Upon recognition of a liability for an uncertain tax position, the Organization would recognize interest expense and penalties in operating expenses.

The Organization files information tax returns in the U.S. federal jurisdiction and the state of Illinois. Its federal and Illinois information tax returns prior to calendar year 2017 are closed. The Organization does not have any tax returns currently under examination by either the Internal Revenue Service (IRS) or any U.S. state jurisdiction.

Membership Dues and Advance Subscriptions Collected

Membership dues and subscriptions are assessed and recognized as revenue based on the life of the dues or subscription.

Functional Allocation of Expenses

The costs of providing the various programs and other activities have been summarized on a functional basis in the statements of activities. The statements of functional expenses present the natural classification detail of expenses by function. Accordingly, certain costs have been allocated among the programs and supporting services benefited.

Expenses which are easily and directly associated with a particular program or supporting service are charged directly to that functional area. Compensation and payroll taxes & fringes have been allocated on the basis of estimates of time and effort.

NOTE 2 – ARRANGEMENTS WITH THE UNIVERSITY OF ILLINOIS

The Organization moved its offices to the University of Illinois at the end of 1999. Under an arrangement with the University, the employees of the Organization are employed by the University. The employees' pension and benefits are part of the University's plans. The Organization then reimburses the University monthly for the cost of its employees.

The University of Illinois holds some cash for the Organization. These "claim on cash" balances were $39,174 and $45,519 as of December 31, 2020 and 2019, respectively. These balances have been included with cash and cash equivalents.

NOTE 3 – CONCENTRATIONS OF CREDIT RISK

Financial instruments that potentially subject the Organization to credit risk consist principally of checking accounts, money markets accounts, and certificates of deposits at financial institutions. However, management continuously monitors the Organization's balances at financial institutions. The balances at each bank as of December 31, 2020 and 2019 were insured by the Federal Deposit Insurance Corporation (FDIC) up to $250,000. As of December 31, 2020, the total bank balance was fully insured. As of December 31, 2019, $4,337 of the bank balance was not insured by FDIC. The "claim on cash" held with the University of Illinois as stated in Note 2 is not insured since it is not held in a financial institution. In addition, the Organization had undeposited funds of $2,124 and $3,309 at December 31, 2020 and 2019, respectively. Since these funds were not yet deposited with a financial institution it was covered by FDIC.

NOTE 4 – INVESTMENTS

FASB Codification 820, Fair Value Measurements, establishes a framework for measuring fair value. That framework provides a fair value hierarchy that prioritizes the inputs to valuation techniques used to measure fair value. The hierarch gives the highest priority to unadjusted quoted prices in active markets for identical assets or liabilities (Level 1 measurements) and the lowest priority to unobservable inputs (Level 3 measurements). The three levels of the fair value hierarchy under FASB Codification 820 are described below:

Level 1 – Inputs to the valuation methodology are based on unadjusted quoted prices for identical assets or liabilities in active markets that the Organization has the ability to access.

Level 2 – Inputs to the valuation methodology include quoted prices for similar assets or liabilities in active markets, quoted prices for identical or similar assets or liabilities in inactive markets, inputs other than quoted prices that are observable for the asset or liability, and inputs that are derived principally from or corroborated by observable market data by correlation or other means. If the asset or liability has a specified (contractual) term, the Level 2 input must be observable for substantially the full term of the asset or liability.

Level 3 – Inputs to the valuation methodology are unobservable and significant to the fair value measurement.

The asset's and liability's fair value measurement level within the fair value hierarchy is based on the lowest level of any input that is significant to the fair value measurement. Valuation techniques used need to maximize the use of observable inputs and minimize the use of unobservable inputs.

The following is a description of the valuation methodologies used for assets measured at fair value. There have been no changes in the methodologies used at December 31, 2020 and 2019.

Intermediate Term Bond Funds:	Valued at the NAV of shares held by the Organization at year-end based on readily determinable fair values, which are published daily and are the basis for current transactions.
Bayer AG (BAYRY) Stock:	Valued at the stated share price at year-end.
Money Market Funds:	Valued at cost plus accrued interest.

The preceding methods described may produce a fair value calculation that may not be indicative of net realizable value or reflective of future fair values. Furthermore, although the Organization believes its valuation methods are appropriate and consistent with other market participants, the use of different methodologies or assumptions to determine the fair value of certain financial instruments could result in different fair value measurements at the reporting date.

The following table sets forth by level, within the fair value hierarchy, the Organization's assets at fair value as of December 31, 2020 and 2019:

Assets at Fair Value as of December 31, 2020

	Level 1	Level 2	Level 3	Total
Without Donor Restrictions				
Bayer AG (BAYRY)	$ 11,872	-	-	$ 11,872
Money Market Funds	425	-	-	425
Designated by the Board				
Intermediate Term Bond Funds	29,834	-	-	29,834
Bayer AG (BAYRY)	47,488	-	-	47,488
Money Market Funds	1,702	-	-	1,702
Total	$ 91,321	-	-	$ 91,321

Assets at Fair Value as of December 31, 2019

	Level 1	Level 2	Level 3	Total
Without Donor Restrictions				
Bayer AG (BAYRY)	$ 8,112	-	-	$ 8,112
Designated by the Board				
Intermediate Term Bond Funds	27,173	-	-	27,173
Bayer AG (BAYRY)	32,448	-	-	32,448
Total	$ 67,733	-	-	$ 67,733

Net investment earnings for the years ended December 31, 2020 and 2019 are summarized as follows:

	2020	2019
Dividends	$ 3,705	$ 730
Capital Gains Distributions	206	-
Net Unrealized Gains (Losses)	(22,303)	3,140
Total Investment Income	$(18,392)	$ 3,870

This investment income, including unrealized gains and losses, are being reported on the statements of activity as investment income.

NOTE 5 – PROPERTY AND EQUIPMENT

As of December 31, 2020, Property and Equipment consists of:

Furniture and Equipment	$ 18,660
Less: Accumulated Depreciation	(15,932)
Property and Equipment, Net	$ 2,728

As of December 31, 2019, Property and Equipment consists of:

Furniture and Equipment	$ 18,660
Less: Accumulated Depreciation	(14,293)
Property and Equipment, Net	$ 4,367

Depreciation expense for the years ended December 31, 2020 and 2019 was $1,639 and $539, respectively.

NOTE 6 – NET ASSETS - BOARD DESIGNATED

Susan C. Eaton Scholar-Practitioner Memorial Fund

The Organization set up a memorial fund in honor of an author of a 1998 "Perspectives on Work" article who died on December 30, 2003. At its June 1, 2004 meeting, the Organization approved the establishment of an annual Susan C. Eaton Scholar-Practitioner award and grant to be paid to one or more qualified scholar researchers in even-numbered years or practitioners in odd-numbered years doing research in the labor and employment relations or related field. The Organization's Executive Board directed that 10% of any gifts received each year are undesignated and can be used for administrative expenses. As of December 31, 2020 and 2019, these designated funds totaled $20,779 and $23,779, respectively.

Kochan-Sleigh Best Dissertation Award Fund

On March 27, 2006, the Organization set up a designated fund with contributions from a member and matching contributions from General Electric. The fund is designated for a minimum of fifteen years and may be used to pay for the $1,000 annual best dissertation award and plaque. After the fifteen-year period, the fund may be continued or the amount remaining in the fund may be undesignated and become available to the Organization for unrestricted purposes. The Organization's Executive Board directed that 10% of any gifts received each year are undesignated and can be used for administrative expenses. As of December 31, 2020 and 2019, these designated funds totaled $37,439 and $38,676, respectively.

Jim Scoville Award Fund

On January 9, 2009, the Organization set up a designated fund with a $10,000 contribution from the University of Minnesota's Industrial Relations Center to honor a member and retiring professor. The fund was established to pay an annual award of $500 and a plaque for best paper on international and comparative employment issues. The Organization's Executive Board directed that 10% of any gifts received each year are undesignated and can be used for administrative expenses. As of December 31, 2020 and 2019, these designated funds totaled $3,296 and $3,924, respectively.

John T. Dunlop Public Policy Fund

On May 8, 2010, the Organization established a designated fund to subsidize a named John T. Dunlop Public Policy Session at future National Policy Forums and/or at the Organization annual meetings. The Organization's Executive Board directed that 10% of any gifts received each year are undesignated and can be used for administrative expenses. This fund will be so named for five or ten years, at which time the Organization's Executive Board will review the fund to determine whether to continue or modify the fund. As of December 31, 2020 and 2019, these designated funds totaled $19,005 and $16,840, respectively.

Gladys and Walter Gershenfeld Publication Fund

On May 8, 2010, the Organization established a designated fund to honor the memory of long-time members and past presidents Walter and Gladys Gershenfeld for the purpose of supporting the Organization's electronic and print publications. The fund will be so named for ten years, at which time the Organization will review the fund's purpose and uses, and whether to continue or modify the fund. The Organization's Executive Board directed that 10% of any gifts received each year are undesignated and can be used for administrative expenses. As of December 31, 2020 and 2019, these designated funds totaled $17,147 and $17,300, respectively.

Quasi-Endowment Fund

During 2019, the Organization established a quasi-endowment fund for the purpose of supporting the Organization's general operations. The Organization's Executive Board directed that 20% of any gifts received each year are undesignated and can be used for administrative expenses. Funds designated by the Board of Directors to function as an endowment are voluntary and may be reversed by the Board of Directors at any time. Accordingly, they are reported as part of the net assets without donor restrictions. The Organization's Board designated endowment net assets exist as a permanent investment pool to enhance and sustain the operations of the Organization so that it can continue to be involved in the full discussion and exchange of ideas between and among its broad constituencies through meetings, publications, and its various electronic listservs and websites. As of December 31, 2020 and 2019, these designated funds totaled $49,190 and $32,448, respectively.

NOTE 7 – NET ASSETS WITH DONOR RESTRICTIONS

Net assets with donor restrictions at December 31, 2020 and 2019 are restricted for the following purposes or period:

	2020	2019
Subject to expenditures for specified purpose:		
Administer the Business of the University Council of Industrial Relations and Human Resources Programs (UCIRHRP)	$ 8,120	$ 7,660
Administer the PhD Student Consortium	14,911	5,851
Total Net Assets with Donor Restrictions	$ 23,031	$ 13,511

NOTE 8 – ENDOWMENT

Board designated net assets at December 31, 2020 consist of an endowment fund established in 2019 for the purpose of supporting operations of the Organization. The Organization has adopted the provisions of FASB ASC 958-205-45, *Not-for-Entities*, including *Net Asset Classification of Funds Subject to an Enacted Version of the Uniform Prudent Management of Institutional Funds Act, and Enhanced Disclosures for All Endowment Fund.* ASC 958 provides guidance on the net asset classification of donor restricted endowment funds for a non-for-profit organization that is subject to an enacted version of the State Prudent Management of Institutional Funds Act (SPMIFA) and also requires disclosures about endowment funds, both donor restricted and board designated endowment funds.

Interpretation of Relevant Law

The Organization has interpreted SPMIFA as requiring the preservation of the fair value of the original gift as of the gift date of the donor-restricted endowment funds absent explicit donor stipulations to the contrary. As a result of this interpretation, the Organization classifies as permanently restricted net assets (a) the original value of gifts donated to the permanent endowment, (b) the original value of subsequent gifts to the permanent endowment, and (c) accumulations to the permanent endowment made in accordance with the direction of the applicable donor gift instrument at the time the accumulation is added to the fund. The remaining portion of the donor-restricted endowment fund that is not classified as permanently restricted net assets is classified as net assets with donor restrictions - restricted for specified purpose until those amounts are appropriated for expenditures by the Organization in a manner consistent with the standard of prudence prescribed by SPMIFA.

Additionally, in accordance with SPMIFA, the Organization considers the following factors in making a determination to appropriate or accumulate donor-restricted endowment funds:

(1) The duration and preservation of the fund
(2) The purpose of the Organization and the donor-restricted endowment fund
(3) General economic conditions
(4) The possible effect of inflation or deflation
(5) The expected total return from income and the appreciation of investments
(6) Other resources of the Organization
(7) The investment policies of the Organization.

Endowment Net Asset Composition and Changes in Net Assets

The composition of endowment net assets and the changes in endowment net assets as of December 31, 2020 are as follows:

	Without Donor Restrictions
Endowment Net Assets, December 31, 2019	$ 32,448
Member Contributions	33,584
Investment Income	(16,842)
Endowment Net Assets, December 31, 2020	$ 49,190

Return Objectives and Risk Parameters

The primary objectives in the investment management for the portfolio assets is to provide ongoing stability and income for operations, preserving gift assets without subjecting them to substantial risk, and providing additional real growth through new gifts. The mix of investments in the portfolio is examined periodically.

Spending Policy

Spending from the fund should not exceed two to five percent annually depending on growth, general economic conditions, and organizational needs. The spending policy should be reviewed every five to ten years.

NOTE 9 – LIQUIDITY AND AVAILABILITY OF FINANCIAL ASSETS

The following reflects the Organization's financial assets as of the statements of financial position date, reduced by amounts not available for general use because of contractual or donor-imposed restrictions within one year of the statement of financial position date. Amounts available include donor restricted amounts that are available for general expenditure in the following year.

	2020	2019
Fiscal Assets at Year-End		
Cash and Cash Equivalents	$ 325,553	$ 283,130
Investments	91,321	67,733
Accounts Receivable, Net	-	-
Less Contractual or Donor-Imposed Restrictions		
Board Designated Funds	(146,856)	(132,967)
Donor Restrictions for Specific Purposes	(23,031)	(13,511)
Financial Assets Available to Meet Cash Needs for General Expenditure Within One Year	$ 246,987	$ 204,385

As part of the Organization's liquidity management, it has a policy to structure its financial assets to be available as its general expenditures, liabilities, and other obligations come due. Excess cash is generally held in checking accounts, money markets accounts, and certificates of deposits until it is required for operational use.

NOTE 10 – COMMITMENTS

The Organization has signed contracts with hotels in Portland and Detroit for the 2020, 2021, and 2022 Annual Meetings. If these contracts were cancelled at December 31, 2020 and 2019, the Organization would have owed $40,437 and $105,798, respectively, to the Portland and Detroit hotels.

NOTE 11 – PENSION PLAN AND RETIREMENT CONTRIBUTIONS

The University of Illinois (the University) contributes to the State Universities Retirement System of Illinois (SURS), a cost-sharing multiple-employer defined benefit plan with a special funding situation whereby the State of Illinois (the State) makes substantially all actuarially determined required contributions on behalf of the participating employers. SURS was established July 21, 1941 to provide retirement annuities and other benefits for staff members and employees of state universities, certain affiliated organizations, and certain other state educational and scientific agencies and for survivors, dependents, and other beneficiaries of such employees. SURS is considered a component unit of the State of Illinois' financial reporting entity and is included in the State's financial reports as a pension trust fund. SURS is governed by Section 5/15, Chapter 40 of the *Illinois Compiled Statutes*. SURS issues a publicly available financial report that includes financial statements and required supplementary information. That report may be obtained by accessing the website at www.SURS.org.

Benefits Provided. A traditional benefit plan was established in 1941. Public Act 90-0448 enacted effective January 1, 1998, established an alternative defined benefit program known as the portable benefit package. The traditional and portable plan Tier 1 refers to members that began participation prior to January 1, 2011. Public Act 96-0889 revised the traditional and portable benefit plans for members who begin participation on or after January 1, 2011, and who do not have other eligible Illinois reciprocal system services. The revised plan is referred to as Tier 2. New employees are allowed six months after their date of hire to make an irrevocable election. A summary of the benefit provisions can be found in the SURS' comprehensive annual financial report (CAFR) notes to the financial statements.

Contributions. The State of Illinois is primarily responsible for funding SURS on behalf of the individual employers at an actuarially determined amount. Public Act 88-0593 provides a statutory funding plan consisting of two parts: (i) a ramp-up period from 1996 to 2010 and (ii) a period of contributions equal to a level percentage of the payroll of active members of SURS to reach 90 percent of the total actuarial accrued liability by the end of fiscal year 2045. Employer contributions from "trust, federal, and other funds" are provided under Section 15-155(b) of the Illinois Pension Code and require employers to pay contributions which are sufficient to cover the accruing normal costs on behalf of applicable employees. The employer normal cost was 12.29 percent of employee payroll during the period of July 1, 2018 to June 30, 2019. The employer normal cost was 13.02 percent of employee payroll during the period of July 1, 2019 to June 30, 2020. The employer normal cost was 12.70 percent of employee payroll during the period of July 1, 2020 to June 30, 2021. The normal cost is equal to the value of current year's pension benefit and does not include any allocation for the past unfunded liability or interest on the unfunded liability. Plan members are required to contribute 8.0 percent of their annual covered salary. The contribution requirements of plan members and employers are established and may be amended by the Illinois General Assembly. During the years ended December 31, 2020 and 2019, the Organization reimbursed the University of Illinois $21,561 and $21,137, respectively, for the employer's payments to SURS for eligible employees.

In addition to providing pension benefits, the State of Illinois provides certain health, dental and life insurance benefits to annuitants. This includes annuitants of the Organization. Substantially all State employees, including the Organization's employees, may become eligible for postemployment benefits if they eventually become annuitants. Health and dental benefits include basic benefits for annuitants under the State's self-insurance plan and insurance contracts currently in force. Life insurance benefits for annuitants under age 60 are equal to their annual salary at the time of retirement; life insurance benefits for annuitants age 60 or older are limited to $5,000 per annuitant. Currently, the State does not segregate payments made to annuitants from those made to current employees for health, dental and life insurance benefits. These costs are funded by the State and are not an obligation of the Organization.

Employees of the Organization may also elect to participate in several tax deferred annuity plans and defined contribution plans. These are single employer plans under which benefits are provided to participating employees through contracts issued to each individual. Participation and the level of employee contributions are voluntary. The Organization is not required to make contributions.

LERA gratefully acknowledges the continuing support of its 2020–21 Major Sponsors and Organizational Members

Major Sustaining Sponsors

Kaiser Permanente
Ford Motor Company

Major University Sponsors

Cornell University, School of Industrial and Labor Relations
Florida International University, College of Business
Massachusetts Institute of Technology, Sloan School of Management/IWER
Pennsylvania State University, School of Labor and Employment Relations
Rutgers University, School of Management and Labor Relations
University of Illinois at Urbana-Champaign, School of Labor and Employment Relations

Organizational Members

American Federation of Teachers, AFL-CIO
American Train Dispatchers Association
Berkeley Research Group, Labor and Employment Practice
Black Hills Corporation
Boston University, Questrom School of Business
Cornell University, Scheinman Institute on Conflict Resolution
Cornell University, School of Industrial and Labor Relations
European Trade Union Institute
Federal Mediation and Conciliation Service
Florida International University, College of Business
Ford Motor Company
Guercio & Guercio, LLP
Kaiser Permanente
Massachusetts Institute of Technology, Sloan School of Management
Melbourne Law School, Centre for Employment and Labour Relations
Michigan State University, School of Human Resources and Labor Relations
Pennsylvania State University, School of Labor and Employment Relations
Portland General Electric
Rollins College, Department of Business and Social Entrepreneurship
Rutgers School of Management and Labor Relations
Taiwan Institute of Labor, Occupational Safety and Health
U.S. Department of Agriculture
United Auto Workers International
United Steelworkers of America
University of California, Los Angeles, Anderson School of Management
University of California, Los Angeles, Institute for Research on Labor and Employment
University of California-Berkeley, Center for Labor Research and Education
University of California-Berkeley, Institute for Research on Labor and Employment
University of Illinois at Urbana-Champaign, School of Labor and Employment Relations
University of Minnesota, Center for Human Resources and Labor Studies
University of Toronto, Centre for Industrial Relations and Human Resources
University of Wisconsin School for Workers
W. E. Upjohn Institute for Employment Research

For more information about Organizational Membership with the LERA, please visit the LERA website: https://lera.memberclicks.net/organizational-membership.

ALPHABETICAL LIST OF AUTHORS

Tasneem Omar Ava 9

Matthew Capece 55

Dennis Dabney 53

Adrienne Eaton 1

Jalana Ellis 9

Daniela Febres 9

Stephen Havlovic 7

Eung Il Kim 7

Jeongrock Kim 9

Amanda Klavert 9

Charles G. Smith 7

Kyoko Suzuki 29

You are invited to become a member of

THE LABOR AND EMPLOYMENT RELATIONS ASSOCIATION

The Labor and Employment Relations Association (LERA) was founded in 1947 by a group who felt that the growing field of industrial relations required an association in which professionally minded people from different organizations could meet. It was intended to enable all who were professionally interested in industrial relations to become better acquainted and to keep up to date with the practices and ideas at work in the field. To our knowledge there is no other organization that affords the multiparty exchange of ideas we have experienced over the years–a unique and valuable forum. After 73 years, both our academic and practitioners agree with the conviction of the founders that the encouragement, reporting, and critical discussion of research are essential if our professional field is to advance.

Our local and national membership of 5,000 includes representatives of management, unions, government; practitioners in consulting, arbitration, mediation, and law; and scholars and teachers representing many disciplines in colleges and universities in the United States and Canada, as well as abroad. Libraries and institutions interested in the publications of the Association are also invited to become subscribing members and thereby receive the same member publications. Organizational memberships in the Association are also available.

Membership dues cover publications for 12 months and entitle members to the electronic newsletter, membership directory and proceedings of the annual meeting online, and printed copies of the annual research volume, and the magazine, *Perspectives on Work* (volumes also available online). Additional online resources include Chapter Profiles and an IR/HR Degree Program Listing for the United States, Canada, and Australia.

Dues for the 2022 calendar year are listed below. International memberships require an additional $25.00 per year to cover postage costs.

Regular Member	$199.00
Contributing Member	$310.00
Emeritus Member (10 years+ membership and no longer employed)	$110.00
Student Member (full-time, limited to 4 consecutive years)	$30.00
Library Subscriber	$300.00

Tax-deductible financial contributions to the Association to support its educational activities are always welcome. You can make an online contribution at our website, leraweb.org, and choosing "Donate". If you are not already a member, we invite you to join the LERA by sending your membership application and dues payment. You can also join online at the LERA website, leraweb.org by choosing "Membership", or request an invoice. Inquiries regarding membership, meetings, and publications should be addressed to the LERA office.

LABOR AND EMPLOYMENT RELATIONS ASSOCIATION
University of Illinois at Urbana-Champaign
121 LER Building, 504 E. Armory Avenue
Champaign, IL 61820 USA
Phone: 217-333-0072 LERAoffice@illinois.edu Web: leraweb.org

Also in the
Labor and Employment Relations Association Series

Revaluing Work(ers): Toward a Democratic and Sustainable Future
Edited by Tobias Schulze-Cleven and Todd Vachon...$34.95

Reimagining the Governance of Work and Employment
Edited by Dionne Pohler $34.95

Employment and Disability: Issues, Innovations, and Opportunities
Edited by Susanne Bruyere $34.95

No One Size Fits All: Worker Organization, Policy and Movement in a New Economic Age
Edited by Janice Fine, Linda Burnham, Kate Griffith, Minsun Ji, Victor Narro,
Steven Pitts, and Teyo Reyes...$29.95

The Contradictions of Pension Fund Capitalism
Edited by Kevin Skerrett, Chris Roberts, Johanna Weststar, and Simon Archer $29.95

The Evolving Health Care Landscape: How Employees, Organizations, and Institutions are Adapting and Innovating
Edited by Ariel C. Avgar and Timothy J. Vogus.. $29.95

Inequality, Uncertainty, and Opportunity: The Varied and Growing Role of Finance in Labor Relations
Edited by Christian E. Weller.. ...$29.95

The Disunited States of America: Employment Relations Systems in Conflict
Edited by David Jacobs and Peggy Kahn..$29.95

Collective Bargaining Under Duress: Case Studies of Major U.S. Industries
Edited by Howard Stranger, Ann C. Frost, and Paul F. Clark...$29.95

Public Jobs and Political Agendas: The Public Sector in an Era of Economic Stress
Edited by Daniel J. B. Mitchell...$29.95

Book orders and request for prices and order forms for the above LERA publications should be sent to:

ILR Press, Cornell Univ., P.O. Box 6525, Ithaca, NY 14851
Phone 607-277-2211, Fax 607-277-6292
Web www.cornellpress.cornell.edu
Additional volumes are available through ILR Press.

Information about ordering printed copies of the *Proceedings of the LERA Annual Meeting* can be found online at https://lera.memberclicks.net/publications. Orders for printed copies of *Perspectives on Work* and other LERA publications should be sent to the LERA Office:

LERA, 121 LER Building, 504 E. Armory Ave., MC-504, Champaign, IL 61820
Phone 217-333-0072, Fax 217-265-5130
Email LERAoffice@illinois.edu, Web leraweb.org